Annals of the ICRP

ICRP PUBLICATION 101

Assessing Dose of the Representative Person for the Purpose of Radiation Protection of the Public

and

The Optimisation of Radiological Protection: Broadening the Process

Editor
J. VALENTIN

PUBLISHED FOR

The International Commission on Radiological Protection

by

ELSEVIER

ELSEVIER ICRP Publication 101 Annals of the ICRP

Editorial

Every silver lining has a cloud . . .or?

It is relatively rare for the *Annals of the ICRP* to combine more than one report into a single journal issue. Now that the Commission has decided to do so again for the first time since 2001a,b (in *Supporting Guidance 2*), the most obvious reason is also the least important one, namely the number of pages per issue.

In the present issue, we have two reports, each one of them important, concise – and since each of them is too short to form an issue of the *Annals*, the obvious solution was to print them within the same set of covers. However, the Commission would have merged these reports into a single issue in any case, for very good reasons.

In the first place, they cover related topics that are likely to appeal to the same readership. The first part of the present issue is the report on assessing doses to representative persons. It concerns a classical problem: to ensure that members of the public are adequately protected and licensees are able to demonstrate that they are in compliance with regulations, in spite of the fact that individual members of the public can not and should not be monitored. As first explored and explained in the venerable *Publication 7* (1966), this requires modelling of pathways for the transport of radionuclides in the environment and of habits of people who could be exposed through these pathways, and the present report provides the most up-to-date advice and terminology in this respect.

The report on broadening the process of optimisation that forms the second part of this issue represents a different cut, in that it focuses on a specific step in the process rather than a specific type of exposed person. It discusses the most crucial part of that process, because optimisation is at the very heart of successful radiological protection, and ways of making the process more inclusive and achieve and foster a sense of participation and joint responsibility.

Both of the reports deal with the application of the Commission's Recommendations, thus providing more detailed guidance on how to implement the Commission's System of Radiological Protection in practice, and anybody concerned with the practical aspects of radiological protection of the public will need to be familiar with both reports. They were both conceived in ICRP Committee 4 on the Application of ICRP Recommendations, and they were both produced by Task Groups of that Committee.

Furthermore, the Commission's current review, revision, and re-statement of its fundamental Recommendations is now nearing completion, and together, the present two reports form an important building block underpinning those Recommendations.

Still, there will of course be the odd reader who is specifically interested in one of the reports only and who would have preferred to have them printed separately. As most readers will be well aware, the reports in the *Annals of the ICRP* are all also available as downloadable PDF files at www.sciencedirect.com – and there, the present two reports are posted as separate files, so that users may choose to download one, the other, or both of the reports.

This is not the only advantage of publishing reports on the Internet. In the first place, files are searchable, references are linked when possible directly to the original paper cited, distribution is instantaneous and available around the clock and around the world... and thanks to the far-sighted attitude of our publishers, Elsevier, the set of files will soon include all ICRP reports from day 1 (i.e., from 1928), thus including those earliest reports that were not printed in a publication series of our own but as proffered papers in open scientific journals.

Through Internet publishing, we are also able to reach a much bigger audience than we could ever have hoped for when conventional printing was our only option. Moreover, the Commission and Elsevier proudly participate in the HINARI programme, so that our reports are available totally cost-free in 69 developing countries and at a minimal token cost only in some 30 more countries. Real printed books and journals are still easier to read than electronic copies, and will never go out of fashion as such, but Internet publishing certainly does provide added value.

And this brings us back to the present publication which is thus available as a single printed issue of the *Annals of the ICRP* and as PDF files representing each of the two reports comprising the publication. You actually can both have your cake and eat it!

Jack Valentin

Assessing Dose of the Representative Person for the Purpose of Radiation Protection of the Public

ICRP PUBLICATION 101, PART 1

ELSEVIER

ICRP Publication 101

Annals of the ICRP

Assessing dose of the representative person for the purpose of radiation protection of the public

ICRP Publication 101

Approved by the Commission in September 2005

Abstract–The Commission intended that its revised recommendations should be based on a simple, but widely applicable, system of protection that would clarify its objectives and provide a basis for the more formal systems needed by operating managers and regulators. The recommendations would establish quantified constraints, or limits, on individual dose from specified sources. These dose constraints apply to actual or representative people who encounter occupational, medical, and public exposures. This report updates the previous guidance for estimating dose to the public. Dose to the public cannot be measured directly and, in some cases, it cannot be measured at all. Therefore, for the purpose of protection of the public, it is necessary to characterise an individual, either hypothetical or specific, whose dose can be used for determining compliance with the relevant dose constraint. This individual is defined as the 'representative person'. The Commission's goal of protection of the public is achieved if the relevant dose constraint for this individual for a single source is met and radiological protection is optimised.

This report explains the process of estimating annual dose and recognises that a number of different methods are available for this purpose. These methods range from deterministic calculations to more complex probabilistic techniques. In addition, a mixture of these techniques may be applied. In selecting characteristics of the representative person, three important concepts should be borne in mind: reasonableness, sustainability, and homogeneity. Each concept is explained and examples are provided to illustrate their roles. Doses to the public are prospective (may occur in the future) or retrospective (occurred in the past). Prospective doses are for hypothetical individuals who may or may not exist in the future, while retrospective doses are generally calculated for specific individuals.

The Commission recognises that the level of detail afforded by its provision of dose coefficients for six age categories is not necessary in making prospective assessments of dose, given the inherent uncertainties usually associated with estimating dose to the public and with identification of the representative person. It now recommends the use of three age categories for estimating annual dose to the representative person for prospective assessments. These categories are 0–5 years (infant), 6–15 years (child), and 16–70 years (adult). For practical implementation of this recommendation, dose coefficients and habit data for a 1-year-old infant, a 10-year-old child, and an adult should be used to represent the three age categories.

In a probabilistic assessment of dose, whether from a planned facility or an existing situation, the Commission recommends that the representative person should be defined such that the probability is less than about 5% that a person drawn at random from the population will receive a greater dose. If such an assessment indicates that a few tens of people or more could receive doses above the relevant constraint, the characteristics of these people need to be explored. If, following further analysis, it is shown that doses to a few tens of people are indeed likely to exceed the relevant dose constraint, actions to modify the exposure should be considered.

The Commission recognises the role that stakeholders can play in identifying characteristics of the representative person. Involvement of stakeholders can significantly improve the quality, understanding, and acceptability of the characteristics of the representative person and the resulting estimated dose.

Keywords: Public exposures; Representative person; Critical group; Dose assessment; Uncertainty

CONTENTS

PREFACE

On 20 October 2001, the Main Commission of the International Commission on Radiological Protection (ICRP) approved the formation of a new Task Group, reporting to Committee 4, on the definition of the individual. As stated in the terms of reference, the objective of the Task Group was to develop principles that would assist in defining the individual to be used for estimating dose and determining compliance in the Commission's system of protection. These principles were expected to be important as the Commission's recommendations continued to evolve, in part because those recommendations were expected to give more emphasis to the individual rather than to society as a whole. Demonstration of compliance was also to be addressed. Issues related to the critical group and concepts of uncertainty as related to the individual were to be considered.

This report is the outcome of the Task Group's efforts. It forms one of the supporting documents for the Commission's revised recommendations. The report addresses the areas mentioned above and also several other issues that became evident during the course of the Task Group's work. The guidance in this report builds upon and replaces the concept of the critical group implemented previously by ICRP. It also defines the representative person to be used for determining compliance with dose constraints and limits.

The membership of the Task Group was as follows:

J.E. Till (Chairman)	J.R. Cooper	A.C. McEwan
D. Cancio	T. Kosako	C. Zuur

The corresponding members were:

M.E. Clark	D.A. Cool	K. Ulbak

The Task Group wishes to acknowledge the technical assistance of Dr. Wayne Oatway at the National Radiological Protection Board in the UK for support in making calculations related to age-specific dose. The Task Group also acknowledges Ms. Shawn Mohler for her support in developing the graphics in the report, Mr. George Killough for his support with the statistical issues presented in Annex B, and Ms. Cindy Galvin for her support with the editing of the report.

The Task Group would like to thank those organisations and staff that made facilities and support available for its meetings. These include the National Institute of Radiation Hygiene in Denmark; the Ministry of Housing, Spatial Planning, and Environment in The Netherlands; the National Radiological Protection Board (now the Radiation Protection Division of the Health Protection Agency) in the UK; the Research Centre for Engineering, Environment, and Technology in Spain; the Nuclear Energy Agency in France; and the Department of Energy, the Nuclear Regulatory Commission, and the Centers for Disease Control and Prevention in the USA.

The report was approved by the Commission at its meeting in Geneva in September 2005.

EXECUTIVE SUMMARY

(a) On 20 October 2001, the Main Commission of the International Commission on Radiological Protection (ICRP) approved the formation of a new Task Group on the definition of the individual. The objective of the Task Group was to develop principles that assist in defining the individual whose dose is to be used as the basis for determining compliance with relevant dose constraints for the public. Occupational and medical exposures are not considered in this report.

(b) In normal and existing situations, the dose constraint for the public is specified as an annual dose for regulatory and administrative purposes. The Commission recognises uncertainties in dose assessment to the public and the transient nature of many extreme exposure situations. As a result of this inherent uncertainty, the Commission recognises that in establishing compliance for normal situations, there is a possibility that the dose to some individuals may exceed the dose constraint. Provided that the Commission's recommendations have been met, the probability of the dose to any individual exceeding the relevant constraint will be small.

(c) The Commission recognises three types of exposure situations: normal, existing, and emergency. Furthermore, dose assessments may be prospective or retrospective. Prospective doses are for individuals who may receive the dose in the future, and retrospective doses are doses that have occurred in the past.

(d) Dose assessment can be thought of as a multistage process. The first stage is to obtain information about the source, including data on the types and quantities of radionuclides and radiations emitted. The second stage is to obtain information about the environment, specifically the concentrations of radionuclides in environmental media arising from the source in question. The third stage of the process is to combine concentrations with habit data that are defined by an exposure scenario. The fourth stage is to use coefficients that either relate concentrations in air or soil to external dose rates (external doses), or that convert a unit of intake into dose (internal doses). Dose coefficients are estimated using models of radionuclide behaviour and radiation absorption in the body, and have been derived and published by ICRP. The final stage is to sum the contributions from external and internal dose as appropriate. It is important to recognise that dose assessment is an iterative process. As an aid to clarification, and for intakes of radionuclides in particular, it is useful to consider the stages separately.

(e) It is recognised that variability and uncertainty are inherent in any process of defining individual characteristics and in estimating doses. Variability refers to real and identifiable heterogeneity or diversity in nature. Uncertainty arises from unavoidable limitations in the assessment. Whether doses are estimated by using measurement data, by applying models, or through a combination of measurements and calculations, the variability and uncertainty contribute to a distribution of possible values. The degree of variability and uncertainty is represented by the shape and extent of that distribution. The Commission believes that it is up to the regulatory authority to make the final decision on how to include uncertainties in the estimation of dose for compliance purposes.

(f) The Commission draws a distinction between quantities having values that are measured or estimated and quantities that have values that are selected, either by the Commission or by other organisations. For example, dose constraints, weighting factors, and dose coefficients, when used in the process of assessing compliance and in decision making, are selected as fixed point values and are assumed not to be uncertain. The Commission, however, recognises that there are uncertainties in the models linking detriment to dose. These uncertainties are considered in establishing selected values of quantities, such as limits and constraints.

(g) The Commission recognises that for prospective assessment of dose to the public, the level of detail afforded by its recommendations of dose coefficients for six age categories is not necessary given the uncertainties usually associated with these estimates. Therefore, for the purpose of prospective assessments of continuing exposure, the Commission now recommends that three age categories are sufficient for estimating annual dose to the representative person. These categories are 0–5 years (infant), 6–15 years (child), and 16–70 years (adult). The shorter time period is selected for the 0–5-year age category, when dosimetric characteristics are changing most rapidly, to avoid any unwarranted reduction in the importance attached to doses to younger age groups. Use of these three age categories is judged to be sufficient to characterise the radiological impact of a source and to ensure consideration of younger, more sensitive populations. For practical implementation of this recommendation, dose coefficients and corresponding habit data for a 1-year-old infant, a 10-year-old child, and an adult should be used to represent the three age ranges.

(h) If assessed doses to these age groups include significant contributions from radionuclides known to cause relatively high doses to the fetus or breast-fed infant, and they are approaching the value of the relevant dose constraint, the dose to the fetus or breast-fed infant should be assessed separately to ensure that the quantitative recommendations are respected. In light of the fact that this intake will only be received over a very limited proportion of the individual's lifetime, the Commission considers that an appropriate level of protection can be achieved by comparing the assessed dose to the fetus or breast-fed infant with the dose constraint for members of the public.

(i) Dose to the public cannot be measured directly without considerable difficulty. In most cases, it cannot be measured at all. Therefore, for the purpose of protection of the public, it is necessary to characterise an individual receiving a dose that is representative of the more highly exposed individuals in the population. This individual is defined as the 'representative person'. This term is the equivalent of, and replaces, 'average member of the critical group' described in previous ICRP recommendations.

(j) The Commission's goal is achieved when the value of dose to the representative person is less than the dose constraint, and radiological protection has been optimised.

(k) In considering dose to the representative person, a number of factors should be taken into account: (1) the dose assessment must account for all relevant pathways of exposure; (2) the dose assessment must consider spatial distribution of radionuclides to ensure that the group receiving the highest dose is included in the assessment; (3)

habit data should be based on the group or population exposed and must be reasonable, sustainable, and homogeneous; and (4) dose coefficients have to be applied according to specific age categories. Once these factors are taken into account and depending on the assessment approach employed (deterministic, probabilistic, or a mixture), the representative person can be identified and used to determine compliance.

(l) Dose to the representative person may be calculated using several different approaches that range from simple deterministic to probabilistic methods.

(m) In both cases, appropriate habit data are required. If specific habit data for the exposed population are not available, values may be derived from appropriate national or regional population data. A distribution of these data may be used in probabilistic assessments, or a value on the distribution may be selected for deterministic calculations. Established databases suggest that the 95th percentile of consumption rates for many staple foods tend to exceed the mean values by approximately a factor of 3. The Commission considers that using the 95th percentile of behaviour in deterministic calculations is a cautious assumption for defining an intake rate.

(n) Care should be exercised to avoid selecting extreme percentile values for every variable to prevent excessive conservatism in the assessment. Such a result could lead to a significant and unrealistic overestimation of the dose to the representative person, and may unduly burden the design of medical or other facilities. Taken together, the selection of parameter values must represent a reasonable and sustainable exposure scenario.

(o) Deterministic methods involve the direct multiplication of selected point values of parameters and environmental concentrations. The simplest form of deterministic method is screening, where very conservative assumptions are made to estimate dose using concentrations of radionuclides at the point of discharge to the environment. In some situations, people receiving the higher doses are easily identified because site-specific exposure data are readily available and habit information is known. In other situations, identifying these individuals is an iterative process that considers key pathways of exposure and populations receiving doses from the source. Ultimately, a group is identified that is expected to receive the higher doses. The average characteristics of this group are used to estimate dose to the representative person.

(p) It is also possible to use probabilistic methods to estimate dose. Probabilistic methods combine distributions of parameters into a composite distribution that presents a range of possible doses based on their probability of occurrence. The distribution of dose incorporates: (1) the uncertainty and natural variability in the estimated environmental media concentration (i.e. radionuclide concentration in air, water, soil, and food); and (2) uncertainty in the habit data (i.e. breathing rate, food and water ingestion rates, time spent at various activities).

(q) In a prospective probabilistic assessment of dose to individuals, whether from a planned facility or an existing situation, the Commission recommends that the representative person should be defined such that the probability is less than about 5% that a person drawn at random from the population will receive a greater dose. If such an assessment indicates that a few tens of people or more could receive doses above the relevant constraint, the characteristics of these people need to be explored.

If, following further analysis, it is shown that doses to a few tens of people are indeed likely to exceed the relevant dose constraint, actions to modify the exposure should be considered.

(r) In probabilistic assessments, particular attention should be given to the region and accompanying population where the assessment is being conducted to define the representative person. Care should be used to include all individuals whose dose could possibly be representative of people receiving the higher doses.

(s) For retrospective assessments of dose to specific individuals, either for the purpose of determining compliance for a past period of operation of a facility or an existing situation, the Commission recognises that estimated doses above the dose constraint should be evaluated on a case-by-case basis. In some cases, it may be expected that these doses will only continue for a short time or may never be realised. However, if doses to specific individuals exceed the dose constraint and are expected to continue for a protracted period of time, a decision should be made by the operator and the regulator about whether a reduction in the source is required. Such a situation may warrant additional monitoring to reduce uncertainty in the dose estimate or to verify the magnitude of dose. The above considerations should be separate from any decision regarding whether the previous design or operations were in compliance with their basis of authorisation.

(t) The Commission recognises the role that the public can play in helping to identify and characterise the representative person for radiological protection purposes. The extent of stakeholder involvement will vary between countries and situations. Stakeholders can provide input regarding habit data that are specific to their location. In particular, stakeholders can be helpful in determining reasonableness, sustainability, and homogeneity of data. Collaboration with stakeholders can significantly improve the quality, defensibility, and acceptability of characteristics of the representative person, and also strengthen support from stakeholders in the compliance and decision-making process.

(u) Regardless of the approach taken to determine compliance, the Commission stresses that application of the total system of protection, utilising both compliance with quantitative constraints and optimisation of protection, is necessary for radiological protection.

1. INTRODUCTION

(1) The Commission's system of protection is based upon the principles of quantitative standards of protection, complemented by the requirement to optimise the level of protection achieved. The system is intended to provide an appropriate degree of protection for individuals from the risks associated with exposure to ionising radiation.

(2) The Commission concluded that its revised recommendations (ICRP, 2007) should be based on a simple, but widely applicable, general system of protection that would clarify its objectives and provide a basis for the more formal systems needed by operating managers and regulators. The recommendations establish quantified limits and constraints on an individual's annual dose from specified sources. These restrictions are applied to the exposure of actual or hypothetical individuals. Within this scope, the Commission includes numerical restrictions on the exposure of members of the public.

(3) The Commission has previously used the concept of the critical group for defining those people who receive the highest exposures from a particular source or set of sources of radiation for the purposes of applying its recommendations. The recommendations in this report update the previous guidance for estimating annual dose to the public. Although emphasis in this report is on the prospective exposure situation (i.e. dose to the public in the future), some guidance is also provided on retrospective dose (i.e. dose that has already been received).

(4) The dose[1] from a source received by any particular individual depends upon a number of factors, such as time, location, transport of radionuclides through the environment, and the characteristics of the individual. These characteristics include physiological parameters (e.g. breathing rate), dietary information (e.g. consumption rate of various foods), residence data (e.g. type of dwelling), use of local resources (e.g. agricultural resources), recreational activities (e.g. swimming), and any other individual-specific information that is necessary to estimate annual dose. In the assessment of doses, a specific set of these characteristics is referred to as an 'exposure scenario'. In general, the Commission refers to diet, residence, and other information needed to estimate exposure as 'habit data'.

(5) Section 1 of this report addresses the report's objective, provides background information, and describes fundamental principles and concepts. Section 2 reviews the process for estimating dose to members of the public arising from sources. Section 3 discusses the selection of characteristics for the representative person. Section 4 presents other considerations relevant to the representative person. Annex A provides technical information on the analysis of age categories, and Annex B provides information on the assessment of compliance using probabilistic methods.

[1] Unless otherwise stated in this report, 'dose' is taken to mean 'effective dose', including, as appropriate, the committed dose up to 70 years of age from intakes of radionuclides and the contribution from external irradiation.

1.1. Objective

(6) The objective of this report is to provide guidance on how to assess dose to the individual for the purposes of establishing compliance with the Commission's recommendations for the protection of the public.

(7) This updated guidance is necessary as the Commission's system of protection has continued to evolve and the recommendations of the Commission have become a basic element of regulations in many countries. In addition to this evolution within ICRP, the ability to carry out assessments using more sophisticated computer and software tools has improved significantly over the past two decades. Doses can now more readily be estimated probabilistically, so that a distribution of doses can be developed that includes uncertainties rather than a single point estimate of dose. This report also updates principles necessary to implement the system of protection of ICRP so that it is consistent with methods that are being used to estimate doses to individuals. The report clarifies and elaborates on methods for estimating dose to the public in order to compare estimated doses with dose constraints, optimise protection, and aid in the planning and decision making for emergency situations.

(8) The source and the exposed individual are fundamental elements in each category of exposure, whether occupational, medical, or public. There must be a clear understanding and characterisation of the individual for whom the dose is being assessed. For occupational exposure, which is exposure incurred at work and principally as a result of work, characterisation of the exposed individuals and the sources is generally straightforward. Records exist for these individuals, and their exposures are monitored or assessed individually. Likewise, in medical exposure, which is principally the intentional exposure of people as a part of their own medical diagnosis or treatment, the source and the exposures are usually obvious. Occupational and medical exposures, therefore, are not considered further in this report.

(9) Guidance for the protection of future individuals in the case of disposal of long-lived radioactive waste is provided in *Publication 81* (ICRP, 2000a) and remains valid.

(10) Exposure situations are classified in the revised recommendations (ICRP, 2007) into three broad groups: normal situations, existing situations, and emergency situations. The Commission uses normal situations to address those parts of its scope corresponding to any deliberately introduced or maintained human activity that causes, or potentially causes, radiation exposures. Existing situations are those in which sources already exist; they may have been introduced unintentionally, inappropriately, or as a result of past human activities that have since been abandoned. In many circumstances, existing situations can only be controlled by action to modify exposure pathways. Emergency situations relate to unintended or unexpected events that could result in exposures sufficient to warrant consideration of the introduction of countermeasures. Guidance is provided in Section 2 of this report for each of these three groups.

(11) When protection of the public in different exposure situations is being assessed, doses may be estimated either deterministically or probabilistically. In either case, parameter values involved are uncertain and these uncertainties must be

addressed. In the deterministic approach, a single point estimate of dose is generated. Uncertainties are accounted for by selecting parameter values that will reasonably ensure that the dose is not underestimated. In the probabilistic approach, uncertainties are taken into account by including the range of possible parameter values and developing a distribution of doses.

(12) For determining exposures in an existing situation, it may be possible to use measurement data and other habit data that are specific to the location. These site-specific data may reduce the uncertainties in estimated doses significantly. However, it is also likely that in the case of retrospective dose assessment for public exposure, a distribution of possible doses will result.

1.2. Background

(13) The concept of critical group was first introduced in *Publication 7* (ICRP, 1965) to provide a means for evaluating compliance with the Commission's recommendations. Paragraph 15 of that publication states:

'The presence of a critical nuclide in some critical pathways will not cause the same exposure of each member of the population outside an installation, and preoperational investigations [...] will usually establish the existence of one or two groups of people whose characteristics, e.g. habits, location, or age, cause them to receive doses higher than those received by the rest of the population outside the installation and this requires them to be considered separately, i.e. to be designated as critical. Great judgment is necessary in defining such a group in practice and the following aspects will have to be considered. Some of these are the same as the factors influencing the design of routine surveys and only those concerned with the critical group itself are listed below:

- The location and age distribution of the potentially exposed group
  Dietary habits (e.g. special foodstuffs and amounts consumed)
- Special occupational habits (e.g. the handling of fishing gear)
- The type of dwelling (e.g. shielding characteristics)
- Domestic habits (e.g. time spent indoors, frequency of personal washing, and laundering of clothes)
- Hobbies (e.g. sport fishing or sunbathing)

Such groups in the population may be in the vicinity of the installation or at some distant location; they may include adult males, adult females, pregnant women, and children; they may be individuals who eat foodstuffs prepared in a special way or produced in a particular location; or they may be people in a particular industry...The concept of critical group provides a sound and practical way of complying with the Commission's recommendations concerning members of the public....'

(14) Paragraph 16 of *Publication 7* (ICRP, 1965) continues:

'The critical group should be identified in such a way that it is representative of the more highly exposed individuals in the population and is as homogeneous as practicable with respect to radiation dose, that is, with respect to those factors in Paragraph 15 which affect the dose in the specific case considered.'

(15) Paragraph 17 of *Publication 7* (ICRP, 1965) states:

'Once the critical group has been identified in this way, a suitably representative sample of the group should be selected and studied so as to assess their [sic] actual or potential exposure. The average exposure of such a sample should then be regarded as typical of that of the highly exposed individuals and the Commission's recommendations for the maximum permissible doses for individual members of the public applied to the average. The spread of values in the sample will give some measure of its homogeneity with respect to the characteristics of the individual (such as metabolic rates) which may influence the dose received and which are not measured. These individual differences may tend to increase the spread of the individual doses received within the critical group. It must also be recognised that, outside the critical group, there may be a few individuals whose habits and characteristics are dramatically unconventional. Such peculiarities may sometimes mean that these individuals receive doses somewhat higher than those in the critical group.'

(16) The concept of critical group has continued to be used in ICRP publications and has been widely applied in radiological protection. In Paragraph 67 of *Publication 43* (ICRP, 1985), it is noted:

'In an extreme case it may be convenient to define the critical group in terms of a single hypothetical individual, for example when dealing with conditions well in the future which cannot be characterised in detail. Usually, however, the critical group would not consist of one individual nor would it be very large for then homogeneity would be lost. The size of a critical group will usually be up to a few tens of people. In a few cases, where large populations are uniformly exposed, the critical group may be much larger. This guidance on size has certain implications; for example, in habit surveys it is not necessary to search for the most exposed individual within a critical group in order to base controls on that one person. The results of a habit survey at a particular point in time should be regarded as an indicator of an underlying distribution and the value adopted for the mean should not be unduly influenced by the discovery of one or two individuals with extreme habits.'

(17) The 1990 recommendations in *Publication 60* (ICRP, 1991) state:

'These groups are chosen to be representative of the individuals most highly exposed as a result of the source under review. They are required to be reasonably homogeneous with respect to the characteristics that influence their doses from that source. When this is achieved any individual constraints should be applied to the mean values for the critical group. It is implicit that some members of the critical group will receive doses both above and below the group average.'

(18) The Commission continues to endorse the principles developed in *Publications 7*, *43*, and *60* (ICRP, 1965, 1985, 1991) relating to the selection of individuals for the purpose of assessing compliance with the dose constraint. The purpose of this

report is to clarify and elaborate on the application of these principles by taking into account recent experience and advances in assessing dose to members of the public.

1.3. Fundamental principles and concepts

(19) In normal and existing situations, dose constraints for the public are specified in the form of an annual effective dose for regulatory and administrative purposes. The Commission recognises uncertainties in dose assessment to the public and the transient nature of many extreme exposure situations. As a result of this inherent uncertainty, the Commission recognises that in establishing compliance for normal situations, there is a possibility that the dose to some individuals may exceed the dose constraint. Provided that the Commission's recommendations have been met, the probability of the dose to any individual exceeding the relevant constraint will be small.

(20) The Commission's constraint for the public for normal situations is set, in part, on the basis of exposure situations for individuals that are assumed to continue to occur for a number of years into the future (ICRP, 2007). The population being exposed at any given time is made up of a spectrum of individuals composed of a range of ages, and individuals within the population should be afforded protection as they progress in age over the time that exposures are expected to occur.

(21) In most cases, it is not possible to monitor dose directly to members of the public; rather, monitoring should be focused on concentrations of radionuclides in the environment that may lead to exposure of individuals. Since dose to the public is not being measured directly, it must be estimated using environmental concentrations, appropriate habit data, and applying appropriate dose coefficients in the case of intakes of radionuclides. Methods used to calculate dose range from point value estimates (deterministic) to a distribution of doses (probabilistic). In either case or with the application of a mixture of these methods, decision makers need guidance on how to determine when compliance exists.

(22) Starting in the mid 1980s, ICRP began developing age-dependent dose coefficients for members of the public. A series of publications were issued giving dose coefficients for six age groups, based on reference biokinetic and dosimetric models (ICRP, 1989, 1993, 1995, 1996a,b) These dose coefficients, combined with appropriate habit data, can be used for assessing dose from environmental discharges.

(23) In some situations, such as those existing from an accident or earlier practice, dose to the public can be inferred using environmental concentrations and specific habit data. An example of this is reconstructed doses from the Chernobyl accident (IAEA, 1991). In this case, a distribution of doses was developed that could be related to individuals in the population. Generally, these distributions include a number of doses that lie well beyond those received by most of the population and arise from some extreme values in habit data.

(24) In other situations where proposed releases to the environment are under consideration, assumptions may have to be made about the habits of the exposed individuals.

(25) Therefore, for the purpose of protection of the public, it is necessary to characterise an individual receiving a dose that is representative of the more highly exposed individuals in the population. This individual is defined as the 'representative person'. The dose to this individual is the equivalent of, and replaces, the mean dose in the 'critical group' described in previous ICRP recommendations.

(26) The sections that follow describe fundamental elements of the process of dose assessment. They also explain how the representative person is characterised and identified for making decisions about compliance for normal situations, in planning for emergencies, and for determining other aspects of radiological protection for members of the public.

2. ASSESSMENT OF DOSE

2.1. Purpose of dose assessment

(27) Assessment of dose to the public can be made to determine compliance with the relevant dose constraint, to guide decisions on the level of control of exposure, and to help identify actions to be taken to reduce exposure. For example, in the case of controlled discharges to the environment, the results of the comparison with the dose constraint may determine whether additional effluent control is required. Doses over different periods of time are also estimated to allow for planning in accident situations and to determine the conditions under which countermeasures may be taken in the event of an accident. In addition, doses are estimated in the process of optimisation, where it is not merely sufficient to meet the dose constraint, but also necessary to show that radiological protection has been optimised, taking social and economic factors into account (ICRP, 2007).

(28) The type of assessment conducted, and the degree to which specific information is incorporated, will depend on the purpose. In many circumstances, planning, optimisation, and compliance will require different types of assessment. Planning and optimisation, for example, should consider a variety of exposure circumstances and evaluate where there are opportunities for further protective measures. Compliance assessments, in contrast, are usually designed to demonstrate specifically that predetermined conditions either are, or are not, being met. The remainder of this report focuses on how to demonstrate compliance with the relevant dose constraint as recommended by the Commission.

2.2. Types of dose assessment

(29) The Commission recognises three types of exposure situations: normal situations, existing situations, and emergency situations. Dose assessments may be prospective or retrospective (see Table 2.1). Assessments of annual dose may be categorised as either of these two types, depending on whether the dose is estimated for future years (prospective) or past years (retrospective).

(30) Prospective doses are estimated for individuals whose exposure has not yet occurred, while retrospective doses are generally estimated for groups of individuals that are known to have received exposures. In assessing prospective exposures,

Table 2.1. Examples of dose assessment in different exposure situations

Situation	Type of assessment	
	Prospective	Retrospective
Normal	Determining compliance with the relevant dose constraint	Estimating dose to the public from past operations
Existing	Future prolonged exposures (e.g. after remediation)	Past exposures (e.g. occupancy of contaminated lands)
Emergency	Emergency planning	Actual impacts after emergency

individuals are assumed to exist who possess certain habit characteristics, whether or not those characteristics can be related to specific people.

(31) Prospective assessments are undertaken to estimate future exposures and to show whether a proposed course of action (e.g. the introduction of a new source or the continuation of an existing source) is acceptable and optimised. These assessments have to make assumptions about future conditions. The results of such prospective assessments provide the basis for determining compliance once the source has been introduced.

(32) Prospective assessments are also undertaken to indicate whether a continuing situation will comply with the relevant dose constraint for future years. They may incorporate more detailed information about present site-specific conditions, which may lead to less uncertainty because conditions may be better known than a prospective assessment for the more distant future. When a prospective assessment is to be used specifically for developing authorisation for sources and for demonstrating compliance, the form and scope of the assessment should be specified to correspond with the basis for the requirement.

(33) Prospective assessments are conducted in emergency situations where radioactive materials are released to the environment and the public may be exposed. The assessment uses available field data and measurements, and translates them into estimates of dose for decision makers who provide recommendations for short-term protective actions.

(34) Prospective assessments are also used in the late phase of an emergency response, after the event has been controlled and early protective actions have been implemented. The situation posed by any remaining residual radioactivity is essentially one of continuing exposure and is conceptually the same as an existing exposure.

(35) Finally, prospective assessments may be undertaken to assess an existing situation that was previously unrecognised. They may also be part of the information used to determine whether protective actions should be introduced to reduce exposures.

(36) Existing situations may require either prospective or retrospective assessments to determine the implications of proposed actions. When such cases have been identified, the assessment provides the basis for understanding future consequences if no actions are taken and for estimating the dose averted if certain actions are implemented. They also provide information that can be communicated to those exposed and the options that may be available.

(37) Retrospective assessments may be undertaken to demonstrate compliance with the relevant dose constraint or could be used as the basis of epidemiological studies (e.g. as in historical dose reconstruction). They generally incorporate more information in calculations than prospective analyses. Additionally, retrospective assessments may be done after the initial phases of an emergency situation to accurately characterise and report the actual impacts and effects of protective actions that may have been undertaken, to provide information to individuals, and to determine if further countermeasures are appropriate.

(38) In emergency situations, there is the potential for relatively high doses to be delivered over relatively short periods of time. In planning for emergencies, prospective assessments may be made by modelling potential source terms and the populations around a particular source, so that preplanned protective measures can be established. These assessments are used to identify individuals and groups that would be subject to dose constraints for actions if the emergency scenario were to occur. Emergency countermeasures are intended to restrict or control the dose to individuals in these short time periods.

(39) Protective actions for acute exposures in emergency situations are often based on protecting specific groups, such as children. In these situations, age-specific habits and age-specific dose coefficients are used to assess the relevant doses and to make decisions on countermeasures. In emergency responses, therefore, information on age groups or populations that were exposed should be included explicitly in the assessment.

2.3. Overview of the dose-assessment process

(40) Dose assessment can be thought of as a multistage process, as demonstrated in Fig. 2.1. The first stage is to obtain information about the source, including data on the types and quantities of radionuclides and radiations emitted. The second stage is to obtain information about the environment, specifically the concentrations of radionuclides in environmental media arising from the source in question. For doses due to external exposures, either the concentrations in air, soil, or water, or the external dose rates are needed. For doses due to internal exposures, it is necessary to know concentrations in food, water, or air that may be taken into the body. The third stage of the process is to combine concentrations with habit data that are selected based on exposure scenarios of the relevant person or group. For external exposures, the amount of time spent in different radiation fields is needed, while for internal exposures, information on the amount of food and water consumed or air breathed is required to estimate activity intakes. The next stage is to use dose coefficients that either relate concentrations in air or soil to external exposure rates, or that convert a unit of intake into dose. The final stage is to sum the contributions from external and internal exposure as appropriate. It is useful to consider the stages separately.

(41) In the first stage, the source of the exposure should be characterised. In the case of discharges to the environment, this characterisation should include discharges for radionuclides of interest, stack heights, proximities of relevant neighbouring buildings, physical and chemical forms of the material, and meteorological conditions. Direct external exposure from sources through shielding, or via scattering or refraction by material in the atmosphere, should also be examined.

(42) In the second stage, environmental concentrations at various locations are obtained by measurements, by modelling the dispersion, deposition, and transport of radionuclides through environmental media, or by a combination of both. Both measurements and modelling will have associated uncertainties. The result for each location is a distribution of concentrations of activity for each radionuclide and

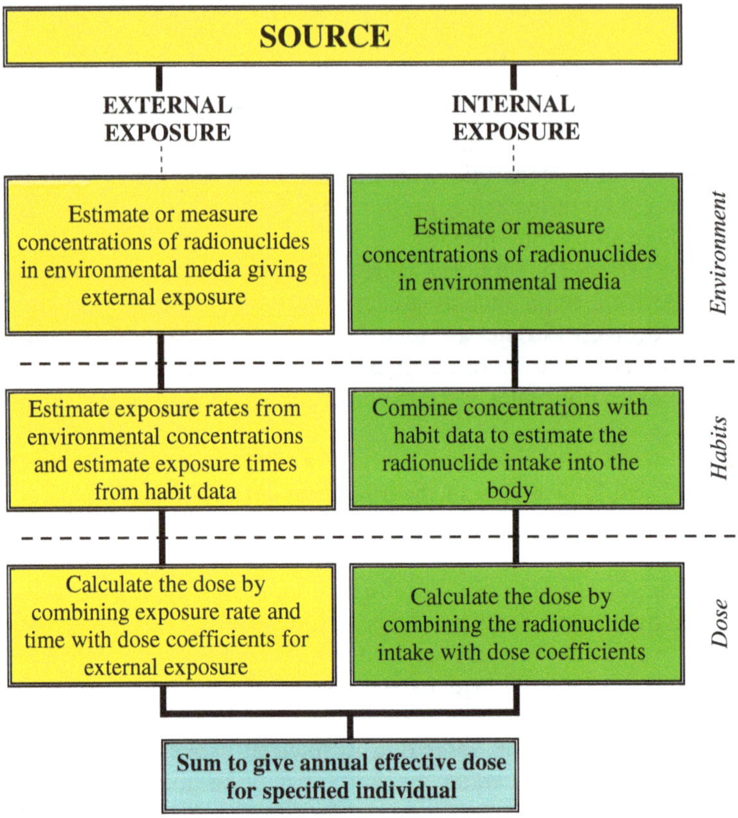

Fig. 2.1. Dose-assessment process.

environmental pathway, as a result of the source. In this stage, the development of the distribution should be independent of the presence or absence of individuals, and it should be based on whether there is a potential pathway of exposure.

(43) The third stage of the process is the combination of concentrations of radionuclides in environmental media with habit data and other information defined by exposure scenarios. Information to be considered includes location, diet, lifestyle activities leading to radiation exposure, and age-dependent physiological factors such as age and breathing rates. The selection of this information is discussed in detail in Section 3. In many cases, these data can be obtained from available information about local populations. However, some situations may require appropriate national or regional information to be used either in the absence of site-specific data, or to supplement or validate such data.

(44) The fourth stage of the dose-assessment process is the application of dose coefficients and related quantities. For intakes of radionuclides, these dose coefficients are expressed in terms of committed effective dose or equivalent dose to an

organ. Summation of the contributions from internal and external exposure results in a total annual dose.

(45) It is important to recognise that dose assessment may be an iterative process. The assessment generally begins with more conservative assumptions for sources, parameter values, habit data, and population size. The results from each iteration are used to determine if more site-specific and realistic information is needed. The use of detailed information is particularly important when the magnitude of the doses calculated approaches the relevant constraint.

(46) This report provides guidance primarily on the third and fourth stages of the process.

2.4. Treatment of uncertainties in dose assessment

(47) It is recognised that variability and uncertainty are inherent in any process of defining individual characteristics and in estimating doses. Variability refers to real and identifiable heterogeneity or diversity in nature. For example, variability may refer to differences in root uptake factors with changing soil types. Sources of variability can be classified into three categories: spatial variability, temporal variability, and interindividual variability (Tschurlovits, 2004). Uncertainty arises from unavoidable limitations in the assessment. For example, measurements of concentrations have inherent limitations in precision. Whether doses are estimated by using measurement data, by applying models, or through a combination of measurements and calculations, the variability and uncertainty contribute to a distribution of possible values. The degree of variability and uncertainty is represented by the shape and extent of that distribution. In this context, sensitivity analysis can be useful for identifying important parameters. This report uses the word 'uncertainty' to refer to the contributions of both variability and uncertainty as described above.

(48) The Commission draws a distinction between quantities having values that are measured or estimated as part of the assessment, and quantities that have values that are selected, either by the Commission or by other organisations. For example, dose constraints, weighting factors, and dose coefficients, when used in the process of assessing compliance and in decision making, are selected as fixed point values and are assumed not to be uncertain. The Commission recognises uncertainties in the models linking detriment to dose. These uncertainties have been taken into account in establishing selected values of quantities such as limits and constraints.

(49) Uncertainties associated with estimation of dose may be taken into account either deterministically by selecting appropriate single values for parameters, or probabilistically by incorporating distributions for parameter values. With either methodology, the goal should be to perform a sufficiently robust evaluation of dose to support the judgments and decisions to be made on radiological protection.

(50) The Commission believes that the final decision on how to include uncertainties in the estimation of dose for compliance purposes should be made by the regulatory authority.

2.5. Deterministic and probabilistic methods for dose assessment

(51) As stated above, dose to the representative person may be calculated either deterministically or probabilistically, or a mixture of these methods may be applied. The method used will depend on the particular situation and the capabilities and data available. Understanding the differences between these methods is important in applying guidance on how compliance with the Commission's recommendations is determined. Therefore, a brief description of these methods follows.

(52) It should be recognised that deterministic and probabilistic methods may not necessarily yield mathematically equivalent results. However, the results of both methods can be used to achieve the Commission's goal of providing a basis for determining the required protection for members of the public. It is important, in both cases, that the assessment process is transparent, assumptions are clearly understood, and the guidance on selection of habits is taken into account. Peer review of the assessment and the involvement of stakeholders are important to the success of the process.

(53) The simplest deterministic method for the assessment of compliance is a screening evaluation. This method typically makes use of simplifying assumptions that lead to a very conservative estimate of dose based on, for example, concentrations of radionuclides at the point of discharge from the source. Another simplifying assumption is to consider a single age group (e.g. adult) in estimating dose to the public to compare with the dose constraints (ICRP, 2000). If the results of relatively conservative screening assessments demonstrate that doses are well below the relevant dose constraint, there may be no need for further detailed assessment of dose. A number of screening methods have been developed and are available for application (IAEA, 2001; NCRP, 1996).

(54) In another form of the deterministic method, a general assessment of the involved populations, pathways, and radionuclides is made with the goal of identifying the group or groups receiving higher doses using expert opinion, measurement data, or simple calculations. In some situations, people receiving the higher doses are easily identified because site-specific exposure data are readily available and habit information is known. In other situations, identifying these individuals is an iterative process that considers key pathways of exposure and populations receiving doses from the source. The iterative process will usually indicate the areas that are likely to receive the greatest exposure from each pathway. These areas should be investigated in more detail. Ultimately, a group is identified that is expected to receive the highest exposure taking all pathways into account. The mean dose to this group is compared with the dose constraint to determine compliance. This method is the same as the critical group approach recommended previously by the Commission.

(55) The propagation of uncertainties through the dose-assessment process is more readily accomplished today than in the past because of advances in computer technology. The probabilistic method combines the distribution of values for parameters into a composite distribution that presents a range of possible doses based on their probability of occurrence. The distribution of calculated dose incorporates: (1) the uncertainty and variability in the estimated environmental media concentration

(i.e. radionuclide concentration in air, water, soil, and food); and (2) uncertainty and variability in the habit data (i.e. breathing rate, food and water ingestion rates, time spent at various activities). As with the deterministic methods, identification of the exposed population and the exposure scenarios of concern are likely to be an iterative process. However, decision makers need guidance on how to determine compliance with the Commission's recommendations when probabilistic methods are used.

(56) A mixture of the deterministic and probabilistic methods is often used. One example of this is the use of measurement data in an existing exposure situation to determine dose to individuals (IAEA, 1991). In this case, a distribution of doses is produced because of the uncertainty and variability of habit and measurement data, and it is this distribution that becomes the basis for determining compliance.

3. THE REPRESENTATIVE PERSON

3.1. Definition of the representative person

(57) Dose to the public must be estimated using environmental concentrations or exposure rates and appropriate habit data. Therefore, for the purpose of protection of the public, it is necessary to define a person to be used for determining compliance with the dose constraint. This is the representative person. This individual, who will almost always be a hypothetical construct, receives a dose that is representative of the more highly exposed individuals in the population. The representative person is equivalent to, and replaces, the average member of the critical group recommended previously by the Commission (ICRP, 1985).

(58) In considering dose to the representative person, a number of factors should be taken into account: (1) the dose assessment must address all relevant pathways of exposure; (2) the dose assessment must consider spatial distribution of radionuclides to ensure that the individuals receiving the higher exposures are included in the assessment; (3) habit data should be based on the population exposed and must be reasonable, sustainable, and homogeneous; and (4) appropriate dose coefficients have to be applied to specific age categories. Once these factors are taken into account, and depending on the assessment approach employed (deterministic, probabilistic, or a mixture of these), the representative person is identified and used to determine compliance. Additional elaboration follows on each of these factors.

3.2. Pathways of exposure, time frames, and spatial distribution of radionuclides

(59) It is important that the dose to the representative person includes appropriate contributions from all modes of exposure (e.g. atmospheric discharges, liquid discharges, and direct external exposure). It is possible that in some assessments, one pathway or a few pathways dominate the exposure. Assumptions can be made so that only the pathways that contribute significantly to the exposure are taken into account. The key to which pathways should be included depends on the type of assessment, but the overall goal should be to ensure that no important pathway has been omitted.

(60) For a time period of about 50 years into the future, it is reasonable to assume that characteristics of individuals can be based on current habit data. The prospective assessment of annual individual dose can therefore be considered valid for a period of this order.

(61) In assessing dose in prospective situations, it may be appropriate to recognise that institutional controls on land use (e.g. designation as a national park or wilderness area) could be in effect. These may preclude types of activity (e.g. residential use or arable cropping) in the designated area so that obtaining staple food supplies from the area would not be possible. Climatic conditions may also preclude or dictate potential for future habitation and locally produced foodstuffs (e.g. in an arid zone, availability of water may preclude both extended residence and sustainable food production). Therefore, the selection of appropriate characteristics should take these conditions into account.

(62) The spatial and temporal distributions of radionuclides discharged and the build-up of long-lived radionuclides over the lifetime of a facility should be taken into account. One example of this build-up is the accumulation in river or lake sediments of radionuclides from liquid releases. Such build-up could result in the most exposed individuals being distant from the facility or being exposed later in time.

(63) The possibility of future changes in land use may need to be considered in a prospective assessment. For example, currently, there may be no agricultural production in the vicinity of a proposed facility, but such production could start during the facility's proposed lifetime. The regulatory authority should determine if this agricultural production is to be assumed in a prospective assessment. Nevertheless, it is important to occasionally re-evaluate the selected characteristics during the lifetime of a facility to account for significant changes that may occur in demographic data and lifestyles.

3.3. Characteristics of the representative person

(64) As indicated in Paragraph 4, characteristics of an individual are described by age-dependent physiological parameters and habit data that include dietary information, residence data, use of local resources, and any other information that is necessary to estimate dose.

(65) It is important that individual habits (e.g. consumption of foodstuffs, breathing rate, location, use of local resources) used in the deterministic approach are average habits of a small number of individuals who are representative of those more highly exposed, and not the extreme habits of a single member of the population. Consideration may be given to some extreme or unusual habits, but they should not necessarily dictate the characteristics of the individuals considered. An exception may be the habits of a single individual that may reasonably be expected to continue for an extended period of time by that individual or others.

(66) When distributions of habit data are employed in a probabilistic approach, the habit data considered should include the range of all possible values found within the relevant population. The distributions of habits should be appropriate for the location or situation under consideration. For example, if discharges to a coastal environment are the subject of an assessment, the distributions of habits should at least reflect the behaviours of residents of coastal communities.

(67) If specific habit data for a local population are not available (e.g. fish consumption from a coastal area with a local discharge of radionuclides into the marine environment), values may be derived from appropriate national or regional population data. A distribution of these data may be used in probabilistic assessments, or a value for habit data on the distribution may be selected for deterministic calculations. Established national databases suggest that the 95th percentile of consumption rates for many staple foods tend to exceed the mean value of the distribution by approximately a factor of 3 (Byrom et al., 1995). The Commission considers that

using the 95th percentile of behaviour in deterministic calculations is a cautious assumption for defining an intake rate in the absence of site-specific data.

(68) Generally, one exposure pathway for a particular source will dominate the dose to the representative person from that source. If more than one intake route for radionuclides provides a significant contribution to dose, it may not be reasonable to assume that the 95th percentile habit data are applicable to all routes; the more dominant route should be assigned a 95th percentile intake, and a lower value should be assigned to other pathways, consistent with the requirement that assessments represent a set of habits that are reasonable and sustainable. Even if more than one exposure pathway makes a significant contribution to the summed effective dose, the individuals receiving the highest exposures tend to be fairly homogeneous with regard to habits (Hunt et al., 1982; Hunt, 2004).

(69) In selecting habit data for the representative person, reasonableness, sustainability, and homogeneity must be considered.

(70) Reasonableness implies that the habit data apply realistically to an individual and are not outside the range of what people encounter in day-to-day life. Reasonableness of habit data must be considered regardless of whether probabilistic or deterministic methods are employed.

(71) Sustainability and homogeneity are aspects of reasonableness. In the deterministic approach, the question of reasonableness in the selection of habit data is related to that of homogeneity because the dose constraint is intended to apply to doses derived from the mean habit data in a reasonably homogeneous group. Homogeneity addresses the degree to which extremes in particular habit data are, or are not, included in the assessment.

(72) For deterministic assessments, the Commission has stated previously (ICRP, 1985) that the necessary degree of homogeneity in habit data in the highest exposed group depends on the magnitude of the mean dose in the group as a fraction of the relevant dose limit or constraint. If that fraction is less than about one-tenth, the group should be regarded as homogenous if the distribution of individual doses lies substantially within a total range of a factor of 10 (i.e. a factor of about 3 on either side of the mean). At fractions above one-tenth, the total range should be less; preferably no more than a factor of 3.

(73) Sustainability addresses the degree to which the selected habits can be continued over the time frame of the assessment. Habit data need to be sustainable. For example, the total dietary intake should be consistent with credible calorific requirements. Habits should correspond to an individual's personal requirements. It is inappropriate to assume, for example, that the same individual receives daily nutrient requirements independently from each of several different pathways (e.g. agriculture and fishing). Also, it is inappropriate to assume that all foods consumed in an area are grown within that area if it is apparent that the location and land area available could not support the assumed dietary intake. Similarly, the intakes of wild game from an area should not exceed feasible game-capture rates. In the case of significant contributions to the dose from external exposure, reasonable estimates of time spent in areas of elevated exposure rates are required. In general,

maintenance of exposure situations for a period of at least 5 years would be considered sustainable.

(74) Care should be exercised to avoid selecting extreme percentile values for every variable to prevent excessive conservatism in the assessment. Such a result could lead to a significant and unrealistic overestimation of the dose to the representative person, and may unduly burden the design of medical or other facilities. Taken together, the selection of parameter values must represent a reasonable and sustainable exposure scenario.

3.4. Age-specific dose coefficients

(75) It is generally recognised that for external exposure in the environment, there is little variability in dose per unit of exposure with age (Golikov et al., 1999, 2000). For intakes of radionuclides, however, the Commission has issued age-specific dose coefficients (dose per unit intake, Sv/Bq) for members of the public in six age ranges covering the time period from infancy to 70 years of age (ICRP, 1996a,b). It has also issued dose coefficients for the embryo/fetus for radionuclide intake by the mother (ICRP, 2001a,b), and dose coefficients for the newborn child for radionuclides in the mother's milk (ICRP, 2005). These coefficients allow the calculation of dose for specific groups in the population. This section provides further guidance on the incorporation of age-specific dose coefficients for internal exposure for the representative person, and distinguishes between their use in different situations. As a basis for understanding the use of age-specific dose coefficients in determining compliance, several goals and fundamental concepts underlying the Commission's recommendations need to be discussed.

(76) The application of dose coefficients for the six age groups should be weighed in relation to the ability to predict concentrations in the environment from a source and the ability to account for uncertainties in habit data for individuals exposed. Uncertainties in estimates of dose, particularly for prospective calculations, are generally not reduced significantly by increasing the number of age categories for which dose coefficients have been provided. The Commission continues to believe that the precision implied by using the full set of dose coefficients is not warranted in the estimation of prospective dose to the public because of the uncertainties involved.

(77) Paragraph 20 points out that the dose constraint is set, at least in part, on the basis of exposures to individuals that are assumed to continue to occur for a number of years into the future. Most facilities are expected to operate for a period of at least 50 years. Therefore, it is the same individual being exposed for a number of years for whom compliance is being determined. This fundamental concept of continuing exposure to the same individual justifies the use of a limited number of age categories that cover several years of a person's life. In the case of disposal of long-lived radioactive waste, where dose to the public may be incurred in the far future over the entire life of the individual, the Commission has stated that '... it is then reasonable to calculate the annual dose/risk averaged over the lifetime of the individuals, which

means that it is not necessary to calculate doses to different age groups; this average can be adequately represented by the annual dose/risk to an adult' (ICRP, 2000).

(78) The dose coefficients provided by the Commission give the committed dose from intake in a single year. This conservative accounting of dose ensures that individuals are protected over a lifetime of exposure, regardless of the number of years for which they are exposed. For example, in the case of actinides, dose coefficients take into account the integrated commitment for a lifetime of exposure, which overestimates dose to an individual in any given year.

(79) The Commission allows averaging over a 5-year period in the evaluation of compliance with the dose constraint (ICRP, 1991), and recommends that a similar approach is appropriate for establishing the number of age groups to be considered in prospective assessments of continuing exposure. Experience to date indicates that age categories can be combined without impacting on protection of members of the public in these situations.

(80) In view of the goals and fundamental concepts underlying the Commission's recommendations as discussed above, some consolidation of the age-specific dose coefficients for internal exposure is warranted. Annex A discusses the implications of such a consolidation. It is evident from these calculations that, with the exception of the actinides, the differences among dose to different age groups are generally small (generally less than a factor of 3) in comparison with uncertainties typically found in assessment of dose to the public.

(81) Therefore, for the purpose of compliance with the dose constraint for continuing exposure, the Commission recommends that the annual dose for the representative person should be defined by three age categories. These categories are 0–5 years (infant), 6–15 years (child), and 16–70 years (adult). The shorter time period is selected for the infant age category, when dosimetric characteristics are changing most rapidly, to avoid any unwarranted reduction in the importance attached to doses to younger age groups. Use of these three age categories is sufficient to characterise the radiological impact of a source and to ensure consideration of younger, more sensitive populations. For practical implementation of this recommendation, dose coefficients and habit data for a 1-year-old (infant), a 10-year-old (child), and an adult should be used to represent the three age categories. These recommendations are summarised in Table 3.1.

(82) The 0–5-year age category does not include the fetus or breast-fed infant. In most cases, the dose to the fetus or breast-fed infant will not be substantially different from the assessed dose for the 0–5-year age category (see Annex A). However, some radionuclides, principally isotopes of phosphorus and the alkaline earths, can deliver significantly higher doses to the fetus and breast-fed infant than to the mother

Table 3.1. Dose coefficients recommended for determining compliance with the dose constraint

Age category (years)	Name of age category	Dose coefficient and habit data to be used
0–5	Infant	1 year old
6–15	Child	10 year old
16–70	Adult	Adult

(ICRP, 2001a,b). Typically, these radionuclides will also deliver relatively higher doses to the infant, so basing compliance on the doses to this age group, using the infant dose coefficient, would normally ensure that the dose to the mother and the fetus is also compliant. Nevertheless, the Commission recognises that the fetus deserves a comparable level of protection. Therefore, if assessed doses to the other age groups include significant contributions from radionuclides known to give rise to relatively high doses to the fetus, and they are approaching the value of the relevant dose constraint, the dose to the fetus or breast-fed infant should be assessed separately to ensure that the quantitative recommendations are respected. In light of the fact that this intake will only be received over a very limited proportion of the individual's lifetime, the Commission considers that an appropriate level of protection can be achieved by comparing the assessed dose to the fetus or breast-fed infant with a dose constraint that could have a higher value than that normally applied to members of the public. The value of the constraint applied to the fetus or breast-fed infant should not, however, exceed the dose limit for members of the public.

(83) This consolidation of age-specific dose coefficients helps provide a system of protection for the public that is robust and allows the continued development of age-specific dosimetry information as the science evolves. The Commission also believes that the use of three age categories is consistent with the derivation of the dose constraint of ICR for the public in normal and existing exposure situations, which are based on continuing exposure of an individual from a source for a number of years.

(84) However, in estimating health effects in retrospective situations, such as dose reconstruction, all of the Commission's age-specific biokinetic models and data continue to be applicable. In these cases, the quality and extent of site-specific data needed to estimate dose generally determine whether age-specific coefficients published by the Commission improve the quality of doses and reduce their uncertainty.

(85) The Commission continues to encourage the use of all available age-specific dose coefficients in planning for and responding to accidents. However, the consolidated age categories proposed by the Commission in this report may be acceptable in some accident situations, especially when prospective assessments are being made of future consequences of the accident or in determining remediation alternatives. This decision should be made by appropriate regulatory authorities.

3.5. Determining compliance

(86) With deterministic methods, the assessment results in a single value of dose that is compared with the relevant constraint to determine compliance. The Commission's goal is achieved when the value of dose to the representative person is less than the dose constraint, and radiological protection has been optimised.

(87) For probabilistic assessments, defining the representative person and determining compliance from a distribution of doses is usually more complex. Annex B describes various distributions of dose and how these distributions may be used to identify the representative person for the purpose of determining compliance. As noted above, there are many acceptable approaches, and numerous distributions

of dose may result. Therefore, the Commission does not prescribe the use of a specific method for probabilistic assessments.

(88) For some probabilistic assessments of dose, it is possible that essentially all doses on the distribution will be predicted to be less than the relevant dose constraint. In this case, compliance is readily demonstrated.

(89) In a prospective probabilistic assessment of dose to individuals, whether from a planned facility or an existing situation, the Commission recommends that the representative person should be defined such that the probability is less than about 5% that a person drawn at random from the population will receive a greater dose.

(90) If such an assessment indicates that a few tens of people or more could receive doses above the relevant constraint, the characteristics of these people need to be explored. A sensitivity analysis of chosen parameter values should be considered to determine if the most appropriate distributions have been used. Attention should also be given to suggestions from members of the public of existing or likely exposure situations that may reflect extremes in the population. Such contributions may not have been included in the operator's analysis. Although those contributions most often correspond to low exposures, experience has shown that they sometimes highlight potentially significant exposure pathways that have not been addressed and that warrant further investigation. If, following further analysis, it is shown that doses to a few tens of people are indeed likely to exceed the relevant dose constraint, actions to modify the exposure should be considered.

(91) For retrospective assessments of dose to specific individuals, either for the purpose of determining compliance for a previous period of operation of a facility or an existing situation, the Commission recognises that estimated doses above the dose constraint should be evaluated on a case-by-case basis. In some cases, it may be expected that these doses will only continue for a short time or may never occur. However, if doses to specific individuals exceed the dose constraint and these doses

Table 3.2. Summary of methods used for determining dose to the representative person

	Calculation method	
	Probabilistic	Deterministic
Environmental concentration data	Distribution of estimated or measured concentration	Single values for parameters
Habit data	Range or fixed values for habit data	Average value for the more highly exposed group or 95th percentile of appropriate national or regional data
Dose coefficient	Fixed value based on age	Fixed value based on age
Dose to the representative person	Method selected by operator or regulator Representative person is identified such that the probability is less than about 5% that a person drawn at random from the population will receive a greater dose	Product of above values

are expected to continue for a protracted period of time, a decision should be made by the operator and the regulator regarding whether a reduction in the source is required. Such a situation may warrant additional monitoring to reduce uncertainty in the dose estimate or to verify the magnitude of dose. The above considerations should be separate from any decision regarding whether the previous design or operations were in compliance with their basis of authorisation.

(92) Regardless of the approach taken to determine compliance, the Commission stresses that application of the total system of protection, utilising both compliance with the relevant constraint and optimisation of protection, is necessary for radiological protection. Table 3.2 summarises the dose-assessment methods described in this section.

4. OTHER CONSIDERATIONS RELEVANT TO THE REPRESENTATIVE PERSON

4.1. Relationship between environmental monitoring, modelling, and the representative person

(93) Prospective assessments are usually undertaken to establish the acceptability of proposed controlled releases to the environment. They almost always involve the use of models, which usually provide the only means of estimating the concentrations of radionuclides in the environmental materials. The representative person should be assumed to occupy a location where the estimated environmental concentrations lead to the higher doses, subject to the requirements for reasonableness, sustainability, and homogeneity noted earlier. Effectively, this means a location that is, or could be, occupied. The temporal and spatial scales represented by the models should also be appropriate for the intended use.

(94) When a source already exists (continuing normal situations, existing situations, and emergency situations), monitoring radionuclide levels in the environment will normally be the most robust method for determining environmental concentrations of radionuclides. This was the primary focus of *Publication 43* (ICRP, 1985). Monitoring programmes should be guided by the identification of dominant pathways and radionuclides, taking the detection limits and source of radionuclides into account. Environmental modelling is also an important and complementary component of monitoring. *Publication 43* (ICRP, 1985) gives guidance on the use and limitations of both monitoring and modelling to estimate doses.

4.2. Situations of potential exposure

(95) In normal exposure situations, annual doses are either being delivered or will certainly occur in the future. However, there may also be situations in which the exposure is not certain to occur and the attributed dose may only have a small probability of occurring. This is termed 'potential exposure'. Potential exposure can cover a wide range of circumstances, including human error, equipment failures, and accidents involving radiation sources, as well as highly non-uniform distribution of radioactive residues.

(96) According to the Commission's recommendations, a potential exposure should be evaluated on the basis of the combination of the probabilities that a radiation dose will be incurred and the probability of harm given that the dose has occurred (ICRP, 1993). The product of these probabilities is the unconditional probability of incurring the health effect. It should be noted, however, that it is more informative for decision-making purposes to present the probability of incurring a dose separately from the dose itself (ICRP, 2000a).

(97) The identification of the representative person in potential exposure situations should take into account the probability of exposure in addition to the other factors in the assessment. The annual dose that such an individual would incur, should the

exposure actually take place, is not a sufficient indicator, although the magnitude of the dose could be important in deciding what risk factor to use. Thus, in addition to characterising the habits, locations, and environmental concentrations of radionuclides, it is necessary to characterise the probability that the individual is exposed. This probability may, in fact, be the combination of several probabilities, including being in the particular location and being engaged in the specific activity that is causing an exposure to occur. In deterministic assessments, this requires selection of the value for probability of exposure to be included in the calculations. For probabilistic assessments, a distribution may be used if such information is available.

(98) An example of an environmental potential exposure is presented by areas contaminated with sparsely distributed hot particles, and has been illustrated in the assessment of the radiological situation created by plutonium hot particles present on the motus (islets) of Colette, Ariel, and Vesta, in the Atoll of Mururoa, French Polynesia (IAEA, 1998). In contrast to the usual assumptions of relatively uniform contamination, there may only be a small probability that the individual would be exposed to one of the hot particles, and an even lower probability that this material would result in an internal exposure through ingestion or absorption through a wound in the body. Even though such a scenario is not likely to occur, should people be exposed to the contaminated area and a hot particle actually be incorporated through a wound, the resulting local dose may be large and could even be a cause of localised deterministic effects such as micronecroses around the incorporated hot particles. The potential for this exposure would remain for as long as the hot particles were present in the environment.

(99) The quantitative approaches to potential exposure are also of use in other situations where exposure may be intermittent or infrequent.

4.3. Value of stakeholder input to characterising the representative person

(100) As noted in *Publication 82* (ICRP, 2000b), the role of stakeholders should be recognised in the wider decision-making process. This role for stakeholders was described in *Publication 82* in the context of protection of the public in situations of prolonged exposure. Since *Publication 82* was published, the Commission has continued to consider and support the role of stakeholders in the system of protection of ICRP. However, the Commission believes that this role needs to be defined, clarified, and expanded to include other situations. Further information on this subject is provided in the following report in this issue of the *Annals of the ICRP*, entitled 'The optimisation of radiological protection: broadening the process'; however, it is important to provide several key principles related to stakeholder involvement in the characterisation of the individual being described in this report.

(101) The decision maker and the operator have clearly defined roles and responsibilities in characterising the individual and in determining compliance. Beyond this, other types of individuals or groups are involoved. They are considered stakeholders, and include individuals who have a personal, financial, health, or legal interest in policy or recommendations that directly affect their well-being or that of their environment. In most cases, the role of stakeholders is to aid and inform the deci-

sion-making process. There may be situations where stakeholders have the authority and responsibility for making or recommending decisions (such as a nationally appointed board or committee). Generally, however, the operator and regulator are the decision makers, and the stakeholders help in the process by providing information and guidance related to decisions being made.

(102) In the case of defining characteristics of the representative person, stakeholder involvement can play an important role. Stakeholders can provide valuable input regarding habit data that are specific to their location. In particular, stakeholders can be helpful in determining the reasonableness, sustainability, and homogeneity of habit data. Collaboration with stakeholders can significantly improve the quality, understanding, and acceptability of characteristics of the representative person, and also strengthen support for the process and the results.

(103) If stakeholder involvement is used as part of the overall decision-making process, guidelines should be established to ensure that the process is effective and meaningful for all parties. Some of these guidelines include, but are not limited to: (1) clear definition of the role of stakeholders at the beginning of the process; (2) agreement on a plan for involvement; (3) provision of a mechanism for documenting and responding to stakeholder involvement; and (4) recognition, by operators and regulators, that stakeholder involvement can be complex and can require additional resources to implement.

(104) The Commission understands that the concept of stakeholder involvement may vary significantly from one country to another for cultural, societal, and political reasons. Therefore, the value and extent of stakeholder involvement should be considered by individual authorities in each country. Nevertheless, the Commission believes that stakeholder involvement can play an important role in the implementation, understanding, and acceptance of the system of protection of ICRP.

ANNEX A: ANALYSIS OF AGE CATEGORIES FOR USE IN ASSESSMENT OF DOSE TO THE PUBLIC

A.1. Introduction

(A1) The Commission has specified dose coefficients for the embryo/fetus, the nursing infant, and for six age groups in the population. This annex investigates whether the number of age groups could be consolidated to three for the purposes of assessing doses for comparison with the relevant dose constraint, particularly in prospective assessments (see Section 3.4). These groups are 0–5 years (infant), 6–15 years (child), and 16–70 years (adult). For practical implementation of this recommendation, dose coefficients and habit data for a 1-year-old infant, a 10-year-old child, and an adult should be used to represent the three age categories.

A.2. Background

(A2) In the mid 1980s, it became apparent that there was a need for dose coefficients (doses per unit intake, Sv/Bq) that could be used for assessing doses from intakes of radionuclides by inhalation and ingestion for all age groups in the population. Task groups of Committee 2 of ICRP developed age-specific biokinetic models that were used for calculating dose coefficients for various ages. In a series of publications (ICRP, 1989, 1993, 1995, 1996a,b), age-specific dose coefficients were given for selected radionuclides of 31 elements. To allow for the effect of body mass and for changes in the biokinetics of radionuclides and dosimetry with increasing age, Committee 2 provided dose coefficients for six representative age groups: the 3-month-old infant; the 1-, 5-, 10-, and 15-year-old child; and the adult. In circumstances for which dose coefficients are required for other age groups in the population, it is recommended that the dose coefficients can be used for the age ranges given below:

- 3-month-old infant, from 0 to 1 year of age;
- 1-year-old child, more than 1 year to 2 years of age;
- 5-year-old child, more than 2 years to 7 years of age;
- 10-year-old child, more than 7 years to 12 years of age;
- 15-year-old child, more than 12 years to 17 years of age; and
- adult, more than 17 years of age.

(A3) Recently, dose coefficients for the embryo and fetus were published in *Publication 88* (ICRP, 2001a,b). Dose coefficients for intakes of radionuclides by infants in their mother's milk have also been developed (ICRP, 2005).

(A4) The dose coefficients published by ICRP for the six age ranges were adopted in the European Basic Safety Standards (EU, 1996) and in the International Basic Safety Standards (IAEA, 1996), as well as being used in many national regulations and guidance notes.

(A5) It is appropriate to use these dose coefficients in circumstances where comprehensive, detailed information is needed on doses to individuals, such as in some dose-reconstruction studies and in dose assessments in planning for and responding to accidents. In many situations, however, this level of detail is not required, and it would be convenient to use a more limited range of coefficients. In deciding whether such a limited range of age-specific dose coefficients is appropriate for use in assessments of dose, it is important to consider the doses that arise from intakes of radionuclides by the different age groups in the population. Comparison of dose coefficients alone is not sufficient. Habit data also have to be taken into account because the radiological criteria against which the results of the assessment would be compared are specified primarily in terms of dose. The intakes of radionuclides by different age groups will not be identical for the same foodstuffs because consumption rates differ among age groups. Therefore, for consumption of the same foodstuff by different age groups, the relative doses will depend not only on the values of the age-specific dose coefficient, but also on the age-specific consumption rates and other biophysical parameters.

(A6) This annex examines the option of using a limited range of age-specific dose coefficients by calculating doses to three selected age groups in the population resulting from ingestion of radionuclides in foods and inhalation of radionuclides in air using representative consumption and breathing rates. These results are compared with doses obtained using all six dose coefficients.

A.3. Method

(A7) For ingestion of foodstuffs, the doses to four age groups (1-, 5-, 10-, and 15-year-old children) were calculated separately using ingestion of unit concentrations of radionuclides in milk, green vegetables, beef, and inhalation of air. The intake rates used are shown in Table A1. These illustrative values are taken from Smith and Jones (2003) and are derived from UK data. Consumption rate data for specific foodstuffs may vary from country to country, but such data usually follow the same general trends. For example, milk is consumed at higher rates by the young, and the rates for solid foods are highest for the adult. Thus, the overall conclusions from this analysis are expected to be generally applicable. Dose calculations were carried out for every radionuclide for which the Commission has published dose coefficients. In

Table A1. Illustrative habit data used in calculating the doses

	Milk consumption rate (kg/year)	Green vegetable consumption rate (kg/year)	Beef consumption rate (kg/year)	Inhalation (m^3/h)
3 months	350	15	10	0.12
1 year	320	30	20	0.22
5 years	280	32.5	25	0.37
10 years	240	35	30	0.64
15 years	260	45	35	0.84
Adult	240	80	45	0.92

addition, the dose to a fetus from the intake of radionuclides by the mother for 9 months, followed by a 3-month period of breastfeeding, was also calculated for those radionuclides where dose coefficients have been published for female members of the public, except where indicated otherwise in Tables A2–A4. For comparison of the inhalation of radionuclides, the dose to a fetus was again calculated for 9 months of exposure from inhalation of radionuclides by the mother, followed by 3 months of inhalation by the offspring at a rate corresponding to a 3-month-old infant and three months of intake by breastfeeding following inhalation by the mother. These calculations were undertaken for a subset of radionuclides that were considered to bound the range of possible results (see Table A6).

Table A2. Ratio of doses from the ingestion of milk using illustrative habit data

Radionuclide	1-year-old child: fetus[†] and 3-month-old breast-fed infant	1-year-old child: 5-year-old child	10-year-old child: 15-year-old child
H-3	2.27	1.77	1.18
H-3 (organically bound tritium)	2.92	1.88	1.25
C-14	3.21	1.85	1.30
Na-22*	6.61	2.04	1.37
Mg-28*	32.05	2.16	1.54
P-32*	1.32	2.31	1.58
S-35 (organic)	5.71	2.29	1.55
K-42*	22.73	2.29	1.47
Ca-45	0.94	2.15	1.28
Fe-59	17.47	1.98	1.40
Co-60	18.70	1.82	1.29
Ni-63	21.02	2.09	1.44
Zn-65	6.51	1.89	1.31
Se-75	6.30	1.79	1.79
Sr-90	2.70	1.78	0.69
Zr-95	25.44	2.13	1.46
Nb-95	15.37	2.03	1.37
Mo-99	13.43	2.22	1.34
Tc-99m	10.67	2.06	1.42
Ru-106	197.98	2.24	1.61
Ag-110m	11.43	2.05	1.41
Sb-125	22.24	2.05	1.38
Te-127m	22.22	2.17	1.60
I-131	7.74	2.06	1.41
Cs-137	3.25	1.43	0.71
Ba-133	15.90	1.82	0.58
Ce-144	2203.39	2.35	1.56
Po-210	56.55	2.29	1.50
Np-237	102.91	1.71	0.92
Pu-239	78.32	1.45	1.04
Am-241	241.09	1.57	1.02
Cm-242	284.84	2.23	1.48

[†] Based on the dose coefficient for a female member of the public for a chronic intake (ICRP, 2001a,b).

* For the fetus calculations, the dose coefficients for these radionuclides are for a working mother (Phipps et al., 2001).

Table A3. Ratio of doses from the ingestion of green vegetables using illustrative habit data

Radionuclide	1-year-old child: fetus[†] and 3-month-old breast-fed infant	1-year-old child: 5- year-old child	10-year-old child: 15-year-old child
H-3	0.64	1.43	0.99
H-3 (organically bound tritium)	0.82	1.52	1.06
C-14	0.90	1.49	1.09
Na-22*	1.86	1.65	1.16
Mg-28*	9.01	1.75	1.30
P-32*	0.37	1.87	1.33
S-35 (organic)	1.61	1.85	1.31
K-42*	6.39	1.85	1.24
Ca-45	0.26	1.74	1.08
Fe-59	4.91	1.60	1.18
Co-60	5.26	1.47	1.08
Ni-63	5.91	1.69	1.21
Zn-65	1.83	1.52	1.11
Se-75	1.77	1.45	1.51
Sr-90	0.76	1.43	0.58
Zr-95	7.15	1.72	1.23
Nb-95	4.32	1.64	1.16
Mo-99	3.78	1.79	1.13
Tc-99m	3.00	1.67	1.19
Ru-106	55.68	1.81	1.36
Ag-110m	3.22	1.66	1.19
Sb-125	6.25	1.66	1.17
Te-127m	6.25	1.75	1.35
I-131	2.18	1.66	1.19
Cs-137	0.91	1.15	0.60
Ba-133	4.47	1.47	0.49
Ce-144	619.70	1.89	1.32
Po-210	15.90	1.85	1.26
Np-237	28.94	1.38	0.78
Pu-239	22.03	1.17	0.88
Am-241	67.81	1.26	0.86
Cm-242	80.11	1.80	1.24

[†] Based on the dose coefficient for a female member of the public for a chronic intake (ICRP, 2001a,b).

* For the fetus calculations, the dose coefficients for these radionuclides are for a working mother (Phipps et al., 2001).

A.4. Results

(A8) The results for selected radionuclides for each exposure pathway are given in Tables A2–A6 as ratios. To establish whether the dose to a 1-year-old child can adequately represent the doses for the range of ages from a fetus to 5 years, the ratio of the 1-year-old dose to the fetus/3-month-old breast-fed infant dose, and the ratio of the 1-year-old dose to the 5-year-old dose are provided in the tables. Similarly, to establish whether the dose to a 10-year-old child can adequately represent the range from 6 years to 15 years, the ratio of the 10-year-old dose to the

Table A4. Ratio of doses from the ingestion of beef using illustrative habit data

Radionuclide	1-year-old child: fetus† and 3-month-old breast-fed infant	1-year-old child: 5- year-old child	10-year-old child: 15-year-old child
H-3	0.76	1.24	1.10
H-3 (organically bound tritium)	0.97	1.32	1.16
C-14	1.07	1.29	1.20
Na-22*	2.20	1.43	1.27
Mg-28*	10.68	1.51	1.43
P-32*	20.44	1.62	1.47
S-35 (organic)	1.90	1.60	1.44
K-42*	7.58	1.60	1.37
Ca-45	0.31	1.51	1.19
Fe-59	5.82	1.39	1.30
Co-60	6.23	1.27	1.19
Ni-63	7.01	1.46	1.33
Zn-65	2.17	1.32	1.22
Se-75	2.10	1.25	1.66
Sr-90	0.90	1.24	0.64
Zr-95	8.48	1.49	1.36
Nb-95	5.12	1.42	1.27
Mo-99	4.48	1.56	1.24
Tc-99m	3.56	1.44	1.32
Ru-106	65.99	1.57	1.50
Ag-110m	3.81	1.44	1.31
Sb-125	7.41	1.44	1.29
Te-127m	7.41	1.52	1.49
I-131	2.58	1.44	1.31
Cs-137	1.08	1.00	0.66
Ba-133	5.30	1.27	0.54
Ce-144	734.46	1.64	1.45
Po-210	18.85	1.60	1.39
Np-237	34.30	1.20	0.86
Pu-239	26.11	1.02	0.96
Am-241	80.36	1.10	0.94
Cm-242	94.95	1.56	1.37

† Based on the dose coefficient for a female member of the public for a chronic intake (ICRP, 2001a,b).
* For the fetus calculations, the dose coefficients for these radionuclides are for a working mother (Phipps et al., 2001).

15-year-old dose is also presented. It is evident from these calculations that with the exception of the actinides, the differences among dose to different age groups are generally small (generally less than a factor of 3) in comparison with uncertainties typically found in prospective assessments of dose to the public. For the actinides, dose coefficients take into account the integrated commitment for a lifetime of exposure, which tends to overestimate dose to an individual in any given year. The largest underestimate arising from using the restricted range of age groups was around a factor of 4 for both the inhalation pathway for the fetus/breast-fed infant and the pathway for the ingestion of green vegetables for the same age group.

Table A5. Ratio of doses from inhalation using illustrative habit data

Radionuclide	Lung absorption rate	1-year-old child: 3-month-old infant	1-year-old child: 5-year-old child	10-year-old child: 15-year-old child
H-3 (tritium compounds)	F	1.41	1.08	1.06
H-3 (tritium compounds)	M	1.46	1.15	1.18
H-3 (tritium compounds)	S	1.53	0.94	1.03
H-3 (inhalation of organically bound tritium)		1.83	0.93	1.02
C-14	F	2.01	1.11	1.16
C-14	M	1.46	0.98	0.85
C-14	S	1.64	0.92	0.88
Na-22	F	1.38	1.14	1.22
Mg-28	F	1.63	1.27	1.36
Mg-28	M	1.81	1.22	1.17
P-32	F	1.15	1.39	1.40
P-32	M	1.25	1.11	1.01
S-35 (inhalation of sulphur dioxide)		1.29	1.15	1.23
S-35 (inhalation of carbon disulphide)		1.28	1.19	1.24
K-42	F	1.15	1.35	1.32
Ca-45	F	0.96	1.27	1.00
Ca-45	M	1.34	0.99	0.85
Ca-45	S	1.47	0.99	0.84
Fe-59	F	1.13	1.09	1.23
Fe-59	M	1.32	0.98	0.91
Fe-59	S	1.40	0.95	0.87
Co-60	F	1.41	0.98	1.11
Co-60	M	1.48	0.96	0.95
Co-60	S	1.71	0.87	0.90
Ni-63 (inhalation of nickel carbonyl)		1.54	0.99	1.04
Ni-63	F	1.59	1.08	1.11
Ni-63	M	1.39	1.03	1.01
Ni-63	S	1.64	0.95	1.00
Zn-65	F	1.22	1.04	1.16
Zn-65	M	1.40	1.04	0.96
Zn-65	S	1.62	0.91	0.92
Se-75	F	1.41	1.05	1.59
Se-75	M	1.53	1.07	1.00
Se-75	S	1.54	0.96	0.95
Sr-90	F	0.73	1.00	0.59
Sr-90	M	1.34	1.01	0.78
Sr-90	S	1.75	0.88	0.86
Zr-95	F	1.68	1.02	1.14
Zr-95	M	1.47	0.98	0.88
Zr-95	S	1.45	0.94	0.87
Nb-95	F	1.39	1.15	1.22
Nb-95	M	1.40	1.00	0.88
Nb-95	S	1.40	0.97	0.87

Table A5 (*continued*)

Radionuclide	Lung absorption rate	1-year-old child: 3-month-old infant	1-year-old child: 5-year-old child	10-year-old child: 15-year-old child
Mo-99	F	1.36	1.31	1.38
Mo-99	M	1.34	1.19	1.04
Mo-99	S	1.28	1.19	1.08
Tc-99m	F	1.33	1.26	1.22
Tc-99m	M	1.40	1.15	1.08
Tc-99m	S	1.41	1.14	1.07
Ru-106 (inhalation of ruthenium tetroxide)		1.26	1.07	1.28
Ru-106	F	1.38	1.23	1.33
Ru-106	M	1.44	1.02	1.01
Ru-106	S	1.62	0.98	0.98
Ag-110m	F	1.47	1.11	1.17
Ag-110m	M	1.47	0.98	0.99
Ag-110m	S	1.63	0.94	0.91
Sb-125	F	1.43	1.09	1.17
Sb-125	M	1.47	0.95	0.89
Sb-125	S	1.66	0.94	0.87
Te-127m (inhalation of tellurium vapour)		1.28	1.16	1.25
Te-127m	F	1.22	1.28	1.33
Te-127m	M	1.36	1.03	0.91
Te-127m	S	1.48	0.98	0.89
I-131 (inhalation of methyl iodide)		1.83	1.04	1.17
I-131 (inhalation of elemental iodine vapour)		1.73	1.01	1.18
I-131	F	1.83	1.16	1.32
I-131	M	1.25	1.09	1.05
I-131	S	1.29	1.05	0.91
Cs-137	F	1.13	0.89	0.64
Cs-137	M	1.48	0.96	0.90
Cs-137	S	1.67	0.85	0.87
Ba-133	F	0.75	1.03	0.47
Ba-133	M	1.22	0.93	0.71
Ba-133	S	1.66	0.86	0.90
Ce-144	F	1.38	1.15	1.24
Ce-144	M	1.54	1.08	1.02
Ce-144	S	1.57	0.97	0.96
Po-210	F	1.19	1.30	1.29
Po-210	M	1.34	0.98	0.88
Po-210	S	1.43	0.97	0.88
Np-237	F	1.74	0.92	0.81
Np-237	M	1.67	0.85	0.76
Np-237	S	1.59	0.91	0.82
Pu-239	F	1.75	0.79	0.83
Pu-239	M	1.76	0.76	0.78
Pu-239	S	1.66	0.86	0.85

(*continued on next page*)

Table A5 (*continued*)

Radionuclide	Lung absorption rate	1-year-old child: 3-month-old infant	1-year-old child: 5-year-old child	10-year-old child: 15-year-old child
Am-241	F	1.83	0.89	0.83
Am-241	M	1.73	0.80	0.76
Am-241	S	1.59	0.88	0.85
Cm-242	F	1.43	1.25	1.16
Cm-242	M	1.50	0.97	0.87
Cm-242	S	1.45	0.94	0.86

F, fast. M, moderate. S, slow.

Table A6. Illustrative ratios of doses from inhalation of airborne material during time as a fetus and during breastfeeding with exposure as a 1-year-old child

Radionuclide	Lung absorption class	1-year-old child: fetus and breastfeeding and inhalation*
P-32	F (assumed)	0.22
Ca-45	M	1.01
Sr-90	M	1.73
I-131	F	0.78
Cs-137	F	0.53
Pu-239	M	5.25

* Dose to the fetus following nine months of inhalation by the mother added to the dose to the infant from breastfeeding for 3 months (inhalation by the mother) and 3 months of inhalation by the infant. The dose to the one year old child assumes 12 months of inhalation by the child. F, fast. M, moderate.

A.5. Conclusions

(A9) The results for all radionuclides for which the Commission has published dose coefficients are summarised in Table A7. It can be seen that the ratios are within about a factor of 3–4. Therefore, it can be concluded that, in many situations, the dose calculated for a 1-year-old child can adequately represent doses in the age range from the fetus to the 5-year-old child. Similarly, the dose to a 10-year-old child can adequately represent doses in the age range from 6 years to 15 years.

(A10) The use of a limited set of age-specific dose coefficients representing infant (1-year-old dose coefficient), a child (10-year-old dose coefficient), and an adult is consistent with the likely availability of data on consumption rates. Specific consumption rate data for the six age groups for which the Commission has specified dose coefficients are unlikely to be available in most cases. Data on consumption rates for the three broad categories (infant, child, and adult) are more likely to be available at a national level. Doses for the six age groups, however, may be needed in dose-reconstruction studies and in planning for and responding to accidents.

Table A7. Minimum and maximum ratios for the pathways for all radionuclides using illustrative habit data

Pathway	Ratio	Minimum ratio	Maximum ratio
Ingestion of milk	1-year-old child: fetus and 3-month-old breast-fed infant*	0.94	312,888
	1-year-old child: 5-year-old child	1.14	2.51
	10-year-old child: 15-year-old child	0.49	3.15
Ingestion of green vegetables	1-year-old child: fetus and 3-month-old breast-fed infant*	0.26	88,000
	1-year-old child: 5-year-old child	0.92	2.03
	10-year-old child: 15-year-old child	0.41	2.66
Ingestion of beef	1-year-old child: fetus and 3-month-old breast-fed infant*	0.31	104,296
	1-year-old child: 5-year-old child	0.80	1.76
	10-year-old child: 15-year-old child	0.46	2.93
Inhalation	1-year-old child: 3- month-old infant	0.55	2.51
	1-year-old child: 5-year-old child	0.54	1.65
	10-year-old child: 15-year-old child	0.41	2.84

* For those radionuclides where dose coefficients have been published for female members of the public (ICRP, 2001a,b).

ANNEX B: DETERMINING COMPLIANCE WHEN DOSE TO THE PUBLIC IS ESTIMATED PROBABILISTICALLY

B.1. Introduction

(B1) When dose to the public is estimated probabilistically and uncertainties are taken into account, a distribution of possible doses is the result. It is likely that the use of probabilistic methods will become more frequent in the future as improvements in technology improve the ability to account for uncertainties inherent in any estimation of dose. It is also expected that probabilistic methods will become more widespread as the techniques in their application become more familiar to regulators, operators, and stakeholders. Therefore, the Commission needs to provide guidance on the use of a distribution of dose rather than a deterministic (point) estimate of dose for the purpose of determining compliance. This annex provides a discussion of distributions of dose resulting from probabilistic assessments.

(B2) It is not within the Commission's scope to prescribe how doses may be estimated probabilistically. Many different methods exist. With some distributions of dose, it is possible that doses on the extreme high end of the distribution are well below the dose constraint set by the Commission. In this case, it can be readily determined that compliance exists. However, it is also possible that a probabilistic assessment results in doses on the high end of the distribution that exceed the dose constraint set by ICRP. Further guidance is needed to address this situation to ensure that members of the public are protected in accordance with the Commission's dose constraint.

(B3) In considering probabilistic assessments of dose, it is important to distinguish whether the distribution of dose is retrospective or prospective, and whether the doses are from planned or existing situations. Prospective exposures to individuals may be for planned or existing situations. Retrospective exposures usually apply to specific individuals for existing (or formerly existing) exposure situations. The next section addresses these situations.

B.2. Retrospective and prospective dose

(B4) Methods of performing a risk assessment for a radiological source and interpretation of the results vary according to whether the effects were realised in the past (retrospective assessment) or are contemplated for the future (prospective assessment). A prospective assessment may be applied to a newly designed system, apparatus, or facility, or to future operation of an existing source. Similar mathematical and probabilistic techniques may be used in either temporal frame, but the questions that the analyses seek to answer are usually different.

(B5) Retrospective assessments may apply to past acute exposures or to chronic exposures over an extended period in the past. They may seek to provide dose and risk estimates to support epidemiological investigations in the exposed population. They may also provide information to decision makers concerning remediation of contaminated spaces or compensation of individuals with claims of health or property damage. In such cases, the exposures have already been incurred (and in some cases may be continuing). In principle, the exposed individuals are identifiable and their doses may be reconstructed, although possibly from fragmentary and uncertain information. The exposed individuals are specific individuals, and the resulting estimates may be provided in the context of probabilistic uncertainty analysis. Components of such analysis may include uncertainties concerning sources, transport of effluents through the environment, and concentrations of released radioactive material over time in the environmental media to which people were exposed. Other components of uncertainty may relate to data on individual habits that would affect the estimates (e.g. times spent in contaminated locations, quantities of contaminated food that may have been consumed, and ages at the times of particular exposures). Such information may need to be gathered from surveys designed and analysed by specialists, and uncertainties in the survey results would be based on their analysis. Uncertainties in the doses and risks would be affected by both of these components (source/environment and individual habits).

(B6) Distributions of dose to specific individuals in retrospective assessments serve a number of purposes. Primarily, they are used in epidemiological studies or to inform decision makers whether action is warranted to further reduce exposures if the source continues to exist. Since retrospective doses have already occurred, they are generally not within the scope of the Commission, although the Commission's recommendations may provide a useful guideline for their evaluation.

(B7) Prospective assessments often aim to assist in answering questions related to siting and design of planned facilities and to compliance with regulations. In such cases, the exposed population may be unknown, and the analysis must be based on numerous assumptions. In this connection, the concept of a representative person can provide guidance in preliminary assessments based on tentative design, location, and operational assumptions. Such a representative person would be defined by an exposure scenario based on locations, physical characteristics, and habits that, individually, would be reasonably expected possibilities for some individuals in the exposed population. However, the locations, physical characteristics, and habits of the representative person should not collectively correspond to extreme combinations that are almost certain not to be found in the exposed population. When the analysis is applied to present and future operation of an existing facility, the present exposed population should be known, and it should be possible to investigate the existence of potentially high doses. In this case, too, analysis of the habits of an individual can be adapted to provide first-order operational guidance, subject to adjustment if anomalously exposed groups are identified or if conditions exist that may be expected to produce some doses above the Commission's dose constraint for the public.

(B8) Uncertainty analysis of prospective assessments should be distinguished from its counterpart in retrospective assessments as described in Paragraphs B5 and B6. The representative person for a prospective assessment is not a real member of the exposed population and may not resemble any specific individual, but rather is a mathematical construct for defining a criterion for operational guidance. It is based on assumptions that would correspond to a possible dose that is expected to be high relative to doses estimated for most of the population. It is possible that this calculated dose is not realised at all in the exposed population. Therefore, it is important to distinguish uncertainties associated with the source and the environment from ranges of values assigned by analysts to parameters that define the individual (e.g. breathing rate, age, dietary habits, and fractions of time spent in specified exposure situations).

(B9) Ranges of variation in these variables for the individual do not constitute uncertainty distributions; rather, the variables are parameters to be set by the analyst. When ranges of these variables are combined probabilistically with uncertainty distributions that represent releases from the source, and the transport and concentrations in environmental media of the released radioactivity, interpretation of the composite distribution of artificial ranges and real uncertainties is not straightforward. The individual cannot reasonably be interpreted as a random individual chosen from the affected population. It may seem preferable to specify fixed central or slightly conservative values for the parameters that define the individual, leaving the final uncertainty distribution of dose to primarily reflect the real source and environmental uncertainties. If it is necessary to study the sensitivity of potential dose to the parameters, multiple dose distributions may be generated, with each one corresponding to a parameter point of interest for the individual. Such an exercise may indicate the behaviour of a high percentile (possibly the 95th) of the potential dose distribution, and indicate combinations of the individual's parameters to which the potential dose shows the greatest sensitivity. This type of approach helps avoid confusion of the interpretations of sensitivity to parameters and uncertainty associated with real quantities.

B.3. Distributions related to dose

(B10) In the analysis of dose to individuals and populations, the concept of a distribution arises in two primary contexts, with extension to a third.

- Type 1. When uncertainty is considered in estimates of dose to individuals that are derived from model calculations or contamination measurements, the weighting of a dose distribution is usually interpreted as probability, so that one may make statements such as 'The probability that the annual dose to the specified individual does not exceed 1 mSv is 0.95 (or 95%)'. Such distributions assign probability weight to intervals of dose, and the distribution quantifies the analyst's perception of the uncertainty that affects the estimate. This type of distribution may be useful in connection with the individual, as defined in Paragraph B7.
- Type 2. When (deterministic) point estimates of dose are made for all individuals of an exposed population (or for categories of exposure with numbers of individuals known or estimated for each category), the weighting of each dose interval may be the fraction of the total exposed population receiving a dose within that interval. Such a distribution could be used to estimate the fraction of the exposed population whose annual dose does not exceed some specified level, such as 1

mSv, or it could be used to estimate the annual dose that is not exceeded in 95% of the population (i.e. the 95th percentile of the distribution). Dose distributions of this type could be useful in quantifying dose limitation guidance that is to be applied to the vast majority of the population.

- Type 3. When Types 1 and 2 above are combined, a weighting scheme replaces the point estimate of dose to each individual in Item 2 with a marginal probability distribution that expresses the uncertainty of dose to the individual, given the individual's exposure conditions. The distribution representing the aggregate population would need to be interpreted as probability of dose to a randomly chosen member of the exposed population. The uncertainty distributions associated with individuals would be marginal relative to a rather complex joint distribution of exposure and dose, taking into account relevant correlations for factors such as location and common influence of sources of released radioactivity.

(B11) Distributions are estimated from theory or data, assumed on the basis of experience, or assumed generically (i.e. somewhat arbitrarily but of a form considered to be reasonable). In retrospective studies, where dose has been received by specific people and some records exist for reconstruction, one would expect to have available or to develop a database supporting a Type 2 distribution for information on the population and exposure factors. The difficulty is that information may be fragmentary and uncertain for exposed individuals, and the environmental processes that contributed to the exposures must be reconstructed with some combination of historical data assessment and mathematical modelling. Thus, elements of the Type 3 distribution become important. Where modelling of the source term and environmental transport is used, it may be necessary to introduce uncertainties into the structure and parameters of the models in order to estimate uncertainty propagation into exposure and dose. The mechanism for representing the uncertainties in the source term and environmental transport usually takes the form of probability distributions that substitute for parameters in the models.

(B12) Proposed distributions may be based on measurements (authentic data are used when such measurements exist for relevant times, locations, and processes), but it is often necessary to use surrogate data based on other times or locations. In either case, one has the choice of using an empirical distribution, based directly on a histogram of the data, or a theoretical distribution, which is an idealisation of the histogram, probably represented as a continuous curve. The mathematical form of a theoretical distribution is an assumption based on theory or experience (and usually convenience), possibly supported by a demonstration of its consistency with the relevant data. If the data are too fragmentary to be suggestive of a theoretical distribution type, the assumption is generic; common choices for distributions that represent environmental data are normal and lognormal distributions, but others are possible and may be practical. Sometimes theory is suggestive of the standard distributions in a given case, or a trend of the data may indicate a form of theoretical distribution, even in the absence of a theoretical justification for it, in which case the choice is considered empirical.

(B13) In a prospective assessment of dose to members of an exposed population, the population is usually hypothetical, although it may be based on a real population that exists at the time of the study (although this population may change in unknown ways during the future of the exposure). The purpose could be to assess the effect of the location or design of a proposed power plant or other nuclear facility, or it could be to study the effects of future management of an existing source (e.g. contaminated land). In such a prospective study, the uncertainties related to the source term and environmental transport of released radioactivity (as discussed previously) are applicable, whereas individual detail about members of the population is not available (except as assumptions or extrapolations based on a population existing at the time of the study). The purpose is to ensure that the dose constraint is unlikely to be exceeded.

(B14) To this end, it is important to evaluate and understand exposure scenarios for individuals that would lead to high doses relative to the majority of the population. Limitation of the dose to such a representative person ensures the protection of most of the population. One could consider a Type 1 probability distribution of annual dose to the individual, given the exposure scenario, with uncertainty components due to the source term and environmental transport alone. For example, one may determine by reference to the distribution that the annual dose to the individual would exceed 0.3 mSv with only 2% probability, given the exposure scenario. If the exposure scenario is accepted as being at the upper margin of normal habits and characteristics but not extreme, such a conclusion would imply protection of most of the population. In such an exercise, it would be useful to consider only fixed exposure scenarios (i.e. any parameters such as breathing rates or frequencies associated with habits of the individual should be given fixed but possibly conservative values); attempts to introduce probability distributions into the exposure scenarios and to combine them with uncertainty components associated with the source term and environmental transport may produce results that are more difficult to interpret and possibly misleading.

(B15) A Type 2 distribution of dose (population weighted) may be derived for the hypothetical population of a prospective study, but the information in the distribution is limited by the detail contained in the definition of the population and the methods of estimating dose to different categories of individuals (i.e. exposure scenarios). For example, if one was only considering exposure to airborne radioactivity from a point source, and the spatial distribution of the population was marked out in 1-km radial increments of 16 wind sectors, with the number of individuals residing in each 1 km by 22.5-degree subregion beyond 1 km out to 15 km from the source, then with source-term data (or a model of the release) and an atmospheric transport model, it is possible to estimate a ground-level air concentration of the released radionuclide in each subregion (assuming, for simplicity, that there is only one radionuclide in the release). In the simplest exposure scenarios, it is assumed that there is a uniform average breathing rate, that there is no mobility of population among subregions, and that any difference between indoor and outdoor air concentrations is to be neglected. The subregion point estimates (assumed to be local averages) are the product of the annual release (Bq), the diffusion factor for the centre of the subregion (sometimes called χ/Q or 'Chi-over-Q', s/m^3), the breathing rate (m^3/s), and the dose coefficient for inhalation dose (Sv/Bq). It is then possible to construct the population-weighted distribution by tabulating the estimated subregion averages of dose from smallest to largest, along with the population numbers for the respective subregions. For many purposes, it is useful to normalise the distribution by dividing the tabulated population number for each subregion by the total number of individuals in the exposed population.

(B16) A similar exercise to that described in Paragraph B15 leads to a Type 3 distribution when uncertainties for the source term and χ/Q are considered. The interpretation of such a Type 3 distribution would involve statements such as 'The probability is less than 2% that a person drawn at random from the hypothetical population will receive an annual dose exceeding 0.3 mSv'.

B.4. Specific forms of dose distributions

(B17) Appropriate assumption of a mathematical form for a distribution arising in the context of environmental dose assessment depends on the role that the distribution is intended to play in the analysis. It is emphasised that there is nearly always an element of the analyst's experience and judgment that influences the choice; indeed, the success of the undertaking depends on the availability of experienced and skilled personnel to plan and carry out the quantitative analysis. The subject is embedded in decades of statistical and computational theory and practice, of which no summary can be attempted here.

(B18) At the most specialised level, there are distributions that represent parameters in models of the source term and the environmental transport of the released radionuclides. The complexity of the models usually dictates Monte Carlo methods for simulating the propagation of uncertainties into estimates of concentrations in exposure media (e.g. air, soil, food, water). There can be dozens, if not hundreds, of such parameters; some of which depend on primary or surrogate data from source-term-related processes, on estimates of uncertainty in diffusion model predictions of air concentrations of released materials, or on samples from exploratory wells that monitor groundwater. For some parameters, the literature provides guidance; for others, the analyst should make the case in the context of the study.

(B19) Transport models are often empirically tuned to environmental measurements. In such cases, the parameters may be less literally representative of directly observed quantities. Instead, distributions of the parameters may be inferred by regression methods from residuals that are computed as the difference of model predictions and corresponding measurements of the modelled quantity (e.g. concentration of a radionuclide in air or soil). In such a setting, the residuals (or some transformed version of them) are often (but not always) treated as a sample from a normal distribution with zero mean and variance to be determined by the regression; this choice is sometimes suggested by the theoretical background of the regression method. The process is somewhat complex but the results can be quite powerful and persuasive. The models are often non-linear in the parameters under study, and the subject appears in the literature under the name 'non-linear parameter estimation'.

(B20) Sets of environmental data are often presented in a discussion of an existing facility. It is usual to summarise such a data set with reference to a distribution, from which the data are construed as a random sample, although the actual acquisition procedure may not be consistent with this characterisation. Perhaps the most common assumption for legacy data is that of the normal distribution, and the assumption may be used to present confidence intervals for the mean (which if done by traditional textbook methods would also involve the 'Student' t distribution). The normal distribution necessarily assigns probability symmetrically to semi-infinite negative and positive intervals, and this property can present awkward problems of physical interpretation where physically positive quantities are concerned. It is also the case that histograms of primary data often lack the symmetry that characterises the normal distribution. One approach attempts to get around these problems by using truncated forms of the normal distribution (i.e. one or both tails of the distribution are taken off at specified points). The truncated distribution may give a better fit to the data, but it unfortunately loses much of the mathematical tractability of the untruncated distribution.

(B21) A common theoretical paradigm for skewed distributions is the lognormal distribution. A random variable y is said to be lognormally distributed if its natural logarithm $\ln y$ is normally distributed, and the distribution can be thought of as arising from a transformation of the primary data (all of which must be positive) by taking the natural logarithm and applying the normal distribution. Many skewed distributions, however, are not well fitted by the lognormal distribution, and when the sample size is sufficient, it is sometimes argued that an empirical distribution based directly on the data is the most satisfactory representation. If a distribution represented by a smooth frequency curve is required for mathematical or other convenience, and if direct application of the standard distributions must be ruled out, it is usually possible for an experienced practitioner to fit an empirical frequency curve with the desired properties to the frequency histogram. It is possible that the cumulative representation may be used for fitting.

B.5. Normal distribution and the Central Limit Theorem

(B22) The Central Limit Theorem is usually cited as a principal justification for the ubiquity of the normal distribution in observational science. In a very rough form, the Central Limit Theorem states that under appropriate hypotheses, the sequence of probability distributions of the standardised sums of an infinite sequence of independent random variables $\{x_i\}_{i=1}^{\infty}$ tends to the standard normal distribution:

$$P\left[\frac{\sum_{i=1}^{n}(x_i-\mu_i)}{\sqrt{\sum_{i=1}^{n}\sigma_i^2}}<y\right]\rightarrow\frac{1}{\sqrt{2\pi}}\int_{-\infty}^{y}e^{-t^2/2}\mathrm{d}t\quad\text{as } n\rightarrow\infty$$

where x_i has mean μ_i and standard deviation σ_i. The integral expression after the arrow represents the cumulative probability distribution function of the standard normal distribution, evaluated at y. There are numerous references to studies of hypotheses under which this convergence is shown to be valid. Apart from those restrictions, the random variables $\{x_i\}_{i=1}^{\infty}$ are not required to have any specified form of distribution, nor must all have the same form (Wilks, 1962).

(B23) The Central Limit Theorem is usually invoked to support the claim that sums of environmental random variables, even for relatively small n, are approximately normally distributed. However, the approximation may, in some circumstances, be poor, even for moderately large n. The application of the Central Limit Theorem to the lognormal distribution uses the sequence of logarithmically transformed random variables $\{\ln x_i\}_{i=1}^{\infty}$. Since the lognormal distribution has become commonplace in environmental dose studies, the next section examines some theoretical support for processes that may lead to it in this and related fields.

B.6. Occurrence of lognormal distribution

(B24) Considerable discussion of the origin and applications of the lognormal distribution exists in scientific literature. Aitchison and Brown (1969, Chapter 3) give examples and further references. Eqs. (1)–(4) below are similar to their presentation.

(B25) A basic mathematical model that yields a lognormal distribution is a stochastic process satisfying the equation:

$$X_k - X_{k-1} = \varepsilon_k X_{k-1}, \quad k = 1, 2, \ldots \tag{1}$$

where the ε_k are mutually independent and also independent of the X_k preceding them in the sequence. If the process goes on for n steps, the recursion in Eq. (1) may be solved to get:

$$\begin{aligned} X_n &= (1+\varepsilon_n)X_{n-1} = (1+\varepsilon_n)(1+\varepsilon_{n-1})X_{n-2} = \cdots \\ &= (1+\varepsilon_n)(1+\varepsilon_{n-1})\cdots(1+\varepsilon_1)X_0 \end{aligned} \tag{2}$$

(B26) If the ε_k are sufficiently small in magnitude, the approximation $1+\varepsilon_k \approx e^{\varepsilon_k}$ in Eq. (2) may be used:

$$X_n = X_0 e^{\varepsilon_1} e^{\varepsilon_2} \cdots e^{\varepsilon_n} = X_0 \exp\left(\sum_{k=1}^{n} \varepsilon_k\right) \tag{3}$$

(B27) Taking logarithms in Eq. (3) gives:

$$\ln X_n = \ln X_0 + \sum_{k=1}^{n} \varepsilon_k \tag{4}$$

a sum of independently distributed random variables, which by the Central Limit Theorem is asymptotically normally distributed (i.e. tends to a normal distribution as n tends to infinity, as indicated in the previous section), so that the distribution of X_n approaches lognormality. If the index k marks a series of time steps, the model of Eq. (1) could represent a process of growth of an organism resulting from a variety of independent multiplicative effects represented by the ε_k. In such an interpretation, X_n could represent, for example, the mass, height, or some other physical property of the organism after n steps in its growth. In the abstract, the time step is arbitrary; it could represent seconds or years, depending on the context. The model of Eq. (1) could also represent a sum of money invested at compound interest at a rate per time step that varies with random changes in the economy (e.g. if the index k counts years, the ε_k would represent randomly fluctuating annual interest rates). Forecasts of the value of the investment after, say, $n = 30$ years, would reasonably be modelled as being lognormally distributed. In other applications, the index k could count effects unrelated to time.

(B28) The lognormal distribution has been found to be useful in the field of aerosol physics in applications to particle size. Indeed, in discussions of properties relative to particle size, the lognormal distribution seems to be the generic assumption. Aitchison and Brown (1969) presented a breakage model that seems to be relevant to certain populations of particles. They consider abstract elements (e.g. particles), each having a positive dimension (e.g. mass or effective diameter). The population is subjected to a series of independent operations having the random breakage of the elements as their effect.

(B29) A more specific particle model is considered here; a binary process in which at each breakage stage, each particle is severed into exactly two pieces, one of which is a 'small' part of the original. Assume that the original population of particles is of homogeneous density and that the quantity of interest is the mass of the particle. The 'small' fraction is restricted to no more than a fixed fraction φ, which is less than 0.5 (in this illustration, φ is equal to 0.125). The random value of the small fraction is selected from a uniform distribution of numbers between 0 and φ (excluding zero). The complementary fraction represents the larger part.

(B30) If a particle is chosen at random from breakage stage n, it has a unique ancestry of particles leading back to some particle P_0 in the original population. Thus, its mass can be derived from that of P_0 as a sequence of multiplications by independent random factors:

$$\text{mass}(P_n) = \text{mass}(P_0)\eta_1\eta_2\cdots\eta_n \tag{5}$$

where the η_k are identically and independently distributed,

$$\eta_k = \begin{cases} \zeta \text{ with probability } 0.5 \\ 1 - \zeta \text{ with probability } 0.5 \end{cases}$$

and the random variable ζ is uniformly distributed on the interval [0, 0.125] (zero is excluded, 0.125 is not). The probabilities of 0.5 reflect the fact that a particle in stage k may equally likely be the smaller or the larger part resulting from breakage of the parent (stage k-1) particle. Thus, from Eq. (5):

$$\ln(\text{mass}(P_n)) = \ln(\text{mass}(P_0)) + \ln\eta_0 + \ln\eta_1 + \cdots + \ln\eta_n \tag{6}$$

(B31) The Central Limit Theorem is used to conclude that the logarithm of the mass of P_n is asymptotically normally distributed, so that the mass of P_n is asymptotically lognormally distributed. Fig. B1 shows log-probability distributions for the first five and the tenth stages of breakage, beginning with an initial population of particles with masses selected at random from the uniform distribution over the interval (0,1). The near-linear graph for the tenth stage suggests the approach of the stages of breakage to lognormality. These distributions were obtained by Monte Carlo simulations with the model just described, using 200 trials for each stage.

(B32) Such a model as that illustrated in Fig. B1 may be applicable to weathering of some types of soils.

(B33) Another type of particle model is that of a particle population formed by agglomeration of small particles on to larger condensation nuclei. Such a process could be interpreted in the light of the model of Eq. (1), provided that the incremental particles are sufficiently small relative to the acquiring particles. In such cases, that model would lead to the conclusion of asymptotic lognormality of the population after a sufficient number of acquisitions.

(B34) In reality, there are many effects that depart from the simple models suggested here and that thwart convergence to lognormality. It is sometimes found that a population of particles under investigation is more reasonably viewed as a mixture of two or more distinct subpopulations, which themselves are approximately lognormally distributed, but such as to render the superpopulation non-lognormal. For example, windborne particles may at times consist of sand (diameters of 10 to several hundred micrometres), clay, and silt (diameters of less than one to some tens of micrometres), and fine particles such as combustion products (diameters of a few

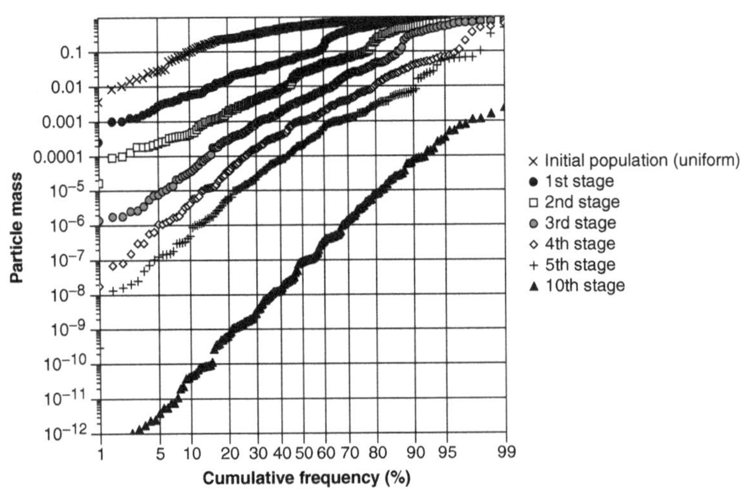

Fig. B1. Sequence of asymmetric binary breakages of particles, beginning with a uniformly distributed population, with mass distribution between 0 and 1. Distributions represent particle masses. The mass of the smaller member of a severed pair has the uniform distribution (0, 0.125). The tendency towards linearity at the tenth stage is suggestive of the effect of the Central Limit Theorem.

hundredths to a few tenths of a micrometre). These distinct subpopulations (which individually are sometimes reasonably approximated as lognormally distributed) may not combine into an aggregate with a distribution that resembles a lognormal. Aerosol scientists often analyse particle size distributions with the lognormal distribution, but there are special cases that are not well fitted by the lognormal distribution and to which the scientists apply special distributions. Some examples are certain coarse dusts (e.g. crushed coal), sprays with large ranges of diameters, and certain powders (Hinds, 1982, Appendix to Chapter 4).

(B35) Other quantities of interest are often sums of more fundamental random variables, which themselves may or may not appear to be lognormally distributed. It is sometimes the case that a sum of lognormally distributed random variables has a distribution that can reasonably be approximated by a lognormal distribution, even though sums of lognormally distributed random variables are known, in general, not to be lognormally distributed. This is a special case of a more general phenomenon, where plotting a dataset of interest suggests that the data resulted from sampling a lognormal distribution, even though no known theory points to such a conclusion. Nevertheless, it is common practice to model skewed empirical distributions generically with the lognormal, even when nothing but experience indicates such a choice. Measurements of radioactivity in environmental media are often construed as lognormal samples with little, if any, theoretical support offered for such an assumption.

(B36) Radiation dose or risk to a specific human population may sometimes fit into the empirically lognormal category just described (e.g. distributions of Type 1, 2, or 3 discussed previously). There are many ways that such a distribution may be arrived at, but few of them obviously imply lognormality, notwithstanding log-probability plots suggesting a linear trend. It is fairly easy to construct simple examples of dose distribution to a population downwind from a release point of airborne radioactivity that are not lognormally distributed with respect to population count (see Example 2 in the next section).

B.7. Examples

(B37) This section develops two relatively simple examples of the distribution types discussed in this annex. Realistic examples would be a great deal more complicated than what is presented here, but the added detail may obscure the main points.

B.7.1. Example 1. Type 1 distribution

(B38) A fictitious existing nuclear facility releases radionuclides to the atmosphere. Studies of data measured at air-monitoring stations and comparisons with release estimates suggest a distribution for the boundary location where annual average air concentrations would be near-maximum. There is theory and evidence to indicate that the concentrations would tend to decrease as the distance from the facility increases. It may be reasonable to position an individual at the maximum boundary location, assuming that he/she lives near that location, spends most days outdoors, and remains near home most of the time. A single released radionuclide is assumed for simplicity. To estimate the dose from inhalation, one would parameterise an equation similar to the following:

$$H_{\text{inhal}} = [\chi/Q]_{\text{annual}} Q_{\text{annual}} (B_{\text{out}} U_{\text{out}} + B_{\text{in}} U_{\text{in}} R) D_{\text{inhal}}$$

where H is the annual dose by inhalation (mSv), $[\chi/Q]$ is the ground-level concentration per unit release rate (s/m^3), D is the dose coefficient for effective dose by inhalation (mSv/Bq), B are breathing rates typical of indoors and outdoors, such as exercising and resting (m^3/s), U are fractions of time spent indoors and outdoors, and R is a fractional factor to estimate a reduced radionuclide concentration in indoor air.

(B39) A similar equation would represent external dose from photons emitted by the airborne radionuclide:

$$H_{\text{extern}} = [\chi/Q]_{\text{annual}} Q_{\text{annual}} (U_{\text{out}} + U_{\text{in}} S) D_{\text{extern}} \times 3.156 \times 10^7$$

where H is the annual external dose from the airborne radionuclide (mSv), $[\chi/Q]$ is the ground-level concentration per unit release rate (s/m^3), D is the dose rate coefficient for effective dose by external exposure to the radionuclide in air (mSv/s/Bq/m^3), S is a fractional factor for lower indoor exposure rate due to reduced indoor concentration and shielding from the outdoor concentration, and 3.156×10^7 is seconds per year.

(B40) It is assumed that all food comes from uncontaminated sources, since there are no known gardens or agricultural operations near the facility.

(B41) Adding the previous equations and re-arranging gives:

$$H_{\text{total}} = H_{\text{inhal}} + H_{\text{extern}} = [\chi/Q]_{\text{annual}} Q_{\text{annual}} K$$

where K is a constant expression that depends on the exposure scenario parameters. The factors for the release Q and atmospheric diffusion $[\chi/Q]$ are subject to uncertainty. It is assumed that the atmospheric data suggest a lognormal distribution and provide an estimate of geometric standard deviation (GSD) of 1.8 for $[\chi/Q]$, and the operator analyses variability in past release data to conclude that Q is lognormally distributed with GSD = 1.3. The product of these independently distributed random variables is lognormally distributed with:

$$\text{GSD} = \exp\sqrt{\ln^2 1.8 + \ln^2 1.3} = 1.9$$

(B42) Lognormality implies that the geometric mean (GM) of H_{total} is the product of the geometric means for $[\chi/Q]$, Q, and the constant K. It is assumed that the data and parameters are such that this product is GM = 0.4 mSv. Fig. B2 is a cumulative log-probability plot of this annual dose distribution. It shows the 95th percentile of the annual dose as 1.15 mSv; thus for this exposure scenario, the probability is 5% that the annual dose could exceed 1.15 mSv. If investigation indicates that few, if any, members of the exposed population are likely to experience this degree of exposure (an assumption that guided the definition of the individual), it is unlikely that members of that population will equal or exceed an annual dose of about 1 mSv from the facility's future releases.

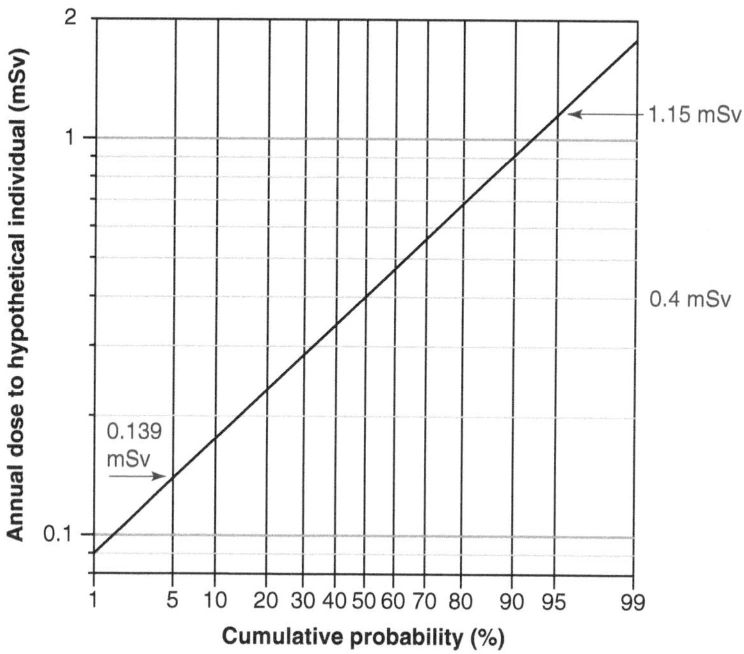

Fig. B2. Type 1 lognormal probability distribution for the individual in Example 1. Given the fixed parameters that define the individual, this distribution assigns probabilities to intervals of dose, based on quantified uncertainties in the release of radionuclides to the atmosphere and variability in observed concentrations at the facility fence near the individual's location. The atmospheric concentrations at this location are assumed to exceed any that the vast majority of the exposed population would encounter. The 95th percentile (1.15 mSv) indicates a probability of 5% that the individual would exceed this annual dose.

(B43) Given the temporal nature of the data for the source term and fence-line air concentrations, it is possible that some would extend the interpretation of the 5% probability to that of exceeding 1.15 mSv in 1 year out of 20. However, one should be cautious of such interpretations for data limited to too few years.

B.7.2. Example 2. Type 2 and Type 3 distributions

(B44) A second fictitious nuclear facility releases radionuclides to the atmosphere (again, for simplicity, one radionuclide is assumed). The exposed population is contained in a single 22.5-degree wind sector from the facility's release stack, between radial distances of 1 km and 15 km from the stack. The spatial distribution of the population is uniform with respect to land area. For the normalised distributions derived here, it is not necessary to know the total number of individuals in the population. The wind sector is divided into 15 subregions, each with a radial distance of 1 km, with population in the 14 subregions that begin at 1 km from the source. The air concentrations of the released radionuclide are estimated with the sector-averaged Gaussian plume model:

$$\chi/Q = \sqrt{\frac{2}{\pi}}\frac{fn}{2\pi\sigma_z ur}\exp(-h^2/(2\sigma_z^2))$$

where f is the fraction of time that the population is downwind from the source, $n = 16$ is the total number of wind sectors (of which only one is used for the example), u is the average wind speed, and r is the radial distance from the source to the point where $[\chi/Q]$ is evaluated. The symbol σ_z (m) denotes a vertical dispersion coefficient for rural Class D conditions (taken as average) given by:

$$\sigma_z = \frac{0.06r}{\sqrt{1 + 0.0015r}}$$

(Hanna et al., 1982). Uncertainties in the predictions of annual average concentrations by this model for regular terrain and meteorological conditions have been estimated by Miller and Hively (1987), and are reasonably interpreted as lognormally distributed with GSD = 1.53 for distances within 10 km and GSD = 2.32 for distances from 10 km to 150 km. Killough and Schmidt (2000) suggested a lognormal distribution with GSD = 1.53 for the uncertainty introduced by the use of composite meteorological data for several recent years. These two uncertainty distributions are used to compute the Type 3 distribution but do not affect the Type 2 distribution.

(B45) For the Type 2 distribution, one computes, deterministically, annual dose as an average for each 1-km subregion. The distribution is plotted in Fig. B3 with dark

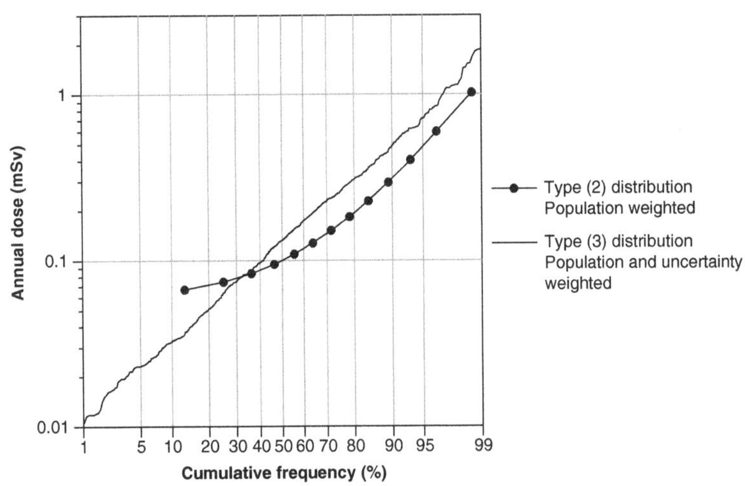

Fig. B3. Distributions of Type 2 and Type 3 in Example 2. The Type 2 distribution weight intervals of deterministically estimated annual dose with fractions of the population receiving those annual doses (note that this distribution lacks the linear trend of lognormality). The Type 3 distribution incorporates uncertainty in the dose estimates, in addition to population proportions, so that the interpretation for a dose interval is the probability that an individual chosen at random from the exposed population will have an annual dose in the interval.

circles connected by line segments. This type of distribution assigns to each annual dose interval the fraction of the population that receives that annual dose. The 50th percentile is 0.1 mSv, and the 95th percentile is about 0.5 mSv. Thus, 5% of this population would receive an annual dose that exceeds 0.5 mSv. There is no implication of quantified uncertainty here, and the interpretation is not probabilistic.

(B46) Computation of the Type 3 distribution requires the uncertainty distributions for the plume model and the meteorological data mentioned previously (a distribution for uncertainty in the annual release would also be required, but that was omitted from the example). The Type 3 distribution assigns intervals of annual dose probabilities based on the uncertainty of the dose estimation. The uncertainty may be associated with the source term and atmospheric transport alone when the population is hypothetical. In retrospective cases, one may also consider uncertainties associated with the population. Fig. B3 shows the example of a Type 3 distribution as the rough line (based on 1000 realisations); the approximate linearity derives from the lognormal distribution that was applied to each individual's dose uncertainty.

(B47) The 95th percentile of the Type 3 distribution is 0.72 mSv, indicating an interpretation that a randomly chosen individual from the exposed population would have a 5% probability of exceeding this annual dose value. However, from the graph, one could estimate a 3% probability of the random individual exceeding an annual dose of 1 mSv. An annual dose in excess of 10×1 mSv (i.e. 10 mSv) is off the graph (well beyond the 99th percentile) and would seem to meet any reasonable definition of 'extremely unlikely'. One should also remember, for example, that a probability of 3% of exceeding 1 mSv given by a Type 3 distribution is not the equivalent of asserting that only 3% of the population would exceed this annual dose. From an inspection of the Type 2 curve (and the output file from the calculations), one can estimate that fewer than 2% of the population would exceed this dose when it is deterministically estimated. In lower-dose regions, the difference between the distributions is greater. For example, the 90th percentile of the Type 3 curve (0.5 mSv) is approximately the same as the (interpolated) 95th percentile of the Type 2 curve.

B.8. Conclusions

(B48) The Commission does not prescribe a specific method to be used for probabilistic assessments. This is because no single mathematical approach or percentage criterion can be applied to the diversity of distributions that may be encountered in probabilistic assessments of dose. Nevertheless, some guidance is necessary to aid operators and regulators in determining when compliance is met when probabilistic assessments are used.

(B49) For some prospective probabilistic assessments of dose, it is possible that essentially all doses on the distribution will be predicted to be less than the dose constraint set by ICRP (e.g. Example 1). In this case, compliance is readily met.

(B50) In a prospective probabilistic assessment of dose to individuals, whether from a planned facility or an existing situation, the Commission recommends that the representative person should be defined such that the probability is less than approximately 5% that a person drawn at random from the population will receive a greater dose. In a large population, many individuals will have doses greater than that of the representative person because of the nature of distributions in probabilistic assessments. This need not be an issue if the doses are less than the relevant dose constraint. However, if such an assessment indicates that a few tens of people or more could receive doses above the relevant constraint, the characteristics of these people need to be explored. If, following further analysis, it is shown that doses to a few tens of people are indeed likely to exceed the relevant dose constraint, actions to modify the exposure should be considered.

(B51) Particular attention should be given to the region and accompanying population where the assessment is being conducted to define the representative person. Care should be taken to include all individuals whose dose could possibly be representative of people receiving higher doses, including extremes. However, it is evident that including too large a region (and population) could dilute the impact of a small number of higher doses and thereby distort the distribution. Therefore, an iterative approach using sequentially smaller regions and populations is generally necessary.

REFERENCES

Aitchison, J., Brown, J.A.C., 1969. The Lognormal Distribution with Special Reference to its Uses in Economics. Cambridge University Press, Cambridge.

Byrom, J., Robinson, C., Simmonds, J.R., Walters, B., Taylor, R.R., 1995. Food consumption rates for use in generalised radiological dose assessments. J. Radiol. Prot. 15, 335–341.

EU, 1996. Council Directive (96/24/EURATOM) of 13 May 1996 Laying Down the Basic Safety Standards for the Protection of the Health of Workers and the General Public Against the Dangers Arising from Ionising Radiation. Off. J. Eur. Commun. L159, 1–29.

Golikov, V., Balonov, M., Erkin, V., Jacob, P., 1999. Model validation for external doses due to environmental contaminations by the Chernobyl accident. Health Phys. 77, 654–661.

Golikov, V., Balonov, M., Jacob, P., 2000. Model of external exposure of population living in the areas contaminated after the Chernobyl accident and its validation. In: Harmonization of Radiation, Human Life and the Ecosystem, Proc. of 10th International Congress of the IRPA, International Conference Centre, Hiroshima, Japan, 746-T-19(1)-2.

Hanna, S.R., Briggs, G.A., Hosker Jr., R.P., 1982. Atmospheric Diffusion Handbook. Report DOE/TIC-11223. US Department of Energy, Washington, DC, USA.

Hinds, W.C., 1982. Aerosol Technology. John Wiley and Sons, New York.

Hunt, G.J., 2004. On homogeneity within the critical group. J. Radiol. Prot. 24, 265–272.

Hunt, G.J., Hewett, C.J., Shepard, J.G., 1982. The identification of critical groups and its application to fish and shellfish consumers in the coastal area of the Northeast Irish Sea. Health Phys. 43, 875–889.

IAEA, 1991. The International Chernobyl Project: Technical Report. Part E, Annex 3: 239. International Atomic Energy Agency, Vienna.

IAEA, 1996. International Basic Safety Standards for Protection against Ionising Radiation and for the Safety of Radioactive Sources. Jointly sponsored by FAO, IAEA, ILO, OECD/NEA, PAHO, WHO, IAEA Safety Series No. 115. International Atomic Energy Agency, Vienna.

IAEA, 1998. To be provided.

IAEA, 2001. Generic Models for Use in Assessing the Impact of Discharges of Radioactive Substances to the Environment. IAEA Safety Report Series No. 19. International Atomic Energy Agency, Vienna.

ICRP, 1965. Principles of Environmental Monitoring Related to the Handling of Radioactive Materials. ICRP Publication 7, Pergamon Press, Oxford.

ICRP, 1966. Principles of Environmental Monitoring Related to the Handling of Radioactive Materials. ICRP Publication 7. Pergamon Press, London, UK.

ICRP, 1985. Principles of monitoring for the radiation protection of the population. ICRP Publication 43, Ann. ICRP 15 (1).

ICRP, 1989. Age-dependent doses to members of the public from intake of radionuclides. Part 1. ICRP Publication 56, Ann. ICRP 20 (2).

ICRP, 1991. 1990 Recommendations of the International Commission on Radiological Protection. ICRP Publication 60, Ann. ICRP 21 (1–3).

ICRP, 1993. Age-dependent doses to members of the public from intake of radionuclides. Part 2. Ingestion dose coefficients. ICRP Publication 67, Ann. ICRP 23 (3/4).

ICRP, 1995. Age-dependent doses to members of the public from intake of radionuclides. Part 3. Ingestion dose coefficients. ICRP Publication 69, Ann. ICRP 25 (1).

ICRP, 1996a. Age-dependent doses to members of the public from intake of radionuclides. Part 4. Inhalation dose coefficients. ICRP Publication 71, Ann. ICRP 25 (3).

ICRP, 1996b. Age-dependent doses to members of the public from intake of radionuclides. Part 5. Compilation of ingestion and inhalation dose coefficients. ICRP Publication 72, Ann. ICRP 26 (1).

ICRP, 2000a. Radiation protection recommendations as applied to the disposal of long-lived solid radioactive waste. ICRP Publication 81, Ann. ICRP 28 (4).

ICRP, 2000b. Protection of the public in situations of prolonged radiation exposure. ICRP Publication 82, Ann. ICRP 29 (1/2).

ICRP, 2001a. Doses to the embryo and fetus from intakes of radionuclides by the mother. ICRP Publication 88, Ann. ICRP 31 (1–3).

ICRP, 2001b. Radiation and your patient: A guide for medical practitioners. Also includes: Diagnostic Reference Levels in medical imaging – review and additional advice. ICRP Supporting Guidance 2, Ann. ICRP 31 (4).

ICRP, 2005. Doses to infants from ingestion of radionuclides in mothers' milk. ICRP Publication 95, Ann. ICRP 34 (3/4).

ICRP, 2007. Recommendations of the International Commission on Radiological Protection. ICRP Publication XX, Ann. ICRP 37 (in press).

Killough, G.G., Schmidt, D.W., 2000. Uncertainty analysis of exposure to radon released from the former feed materials production center. J. Environ. Radioactivity 49, 127–156.

Miller, C.W., Hively, L.M., 1987. A review of validation studies for the Gaussian plume atmospheric dispersion model. Nuclear Safety 28, 522–531.

NCRP, 1996. Screening Models for Releases of Radionuclides to Atmosphere, Surface Water, and Ground. NCRP Report No. 123 (1–2). National Council on Radiation Protection and Measurements, Bethesda, Maryland.

Phipps, A.W., Smith, T.J., Fell, T.P., Harrison, J.D., 2001. Doses to the Embryo/Fetus and Neonate from Intakes of Radionuclides by the Mother. Part 1. Doses Received in Utero and from Activity Present at Birth. HSE Contract Research Report, 397/2001. Health and Safety Executive, London, UK.

Smith, K., Jones, A.L., 2003. Generalised Habit Data for Radiological Assessments. NRPB-W41. National Radiological Protection Board, Chilton, UK.

Tschurlovits, M., Taghizadegan, R., Engelbrecht, R. 2004. Handling Uncertainty and Variability in Risk Communication. Proc. IRPA 11, Madrid. International Radiation Protection Association, www.irpa.net.

Wilks, S.S., 1962. Mathematical Statistics. John Wiley and Sons, Inc, New York.

The Optimisation of Radiological Protection: Broadening the Process

ICRP PUBLICATION 101, PART 2

ELSEVIER

ICRP Publication 101

Annals of the ICRP

The optimisation of radiological protection: Broadening the process

ICRP Publication 101

Approved by the Commision in September 2005

Abstract–The principle of optimisation of radiation protection is defined by the Commission as the source-related process to keep the magnitude of individual doses, the number of people exposed, and the likelihood of potential exposure as low as reasonably achievable below the appropriate dose constraints, with economic and social factors being taken into account. According to the revised recommendations of ICRP, this process of optimisation below constraint should be applied whatever the exposure situation; i.e. planned, emergency, and existing.

The previous recommendations for the practical implementation of the optimisation process are still valid. It must be implemented through an ongoing, cyclical process that involves the evaluation of the exposure situation to identify the need for action, the identification of the possible protective options to keep the exposure as low as reasonably achievable, the selection of the best option under the prevailing circumstances, the implementation of the selected option through an effective optimisation programme, and regular review of the exposure situation to evaluate if the prevailing circumstances call for the implementation of corrective protective actions. However, the way in which the optimisation process should be implemented is now viewed more broadly to reflect the increasing role of individual equity, safety culture, and stakeholder involvement in our modern societies.

This report is a consolidation and an evolution of the Commission's recommendations concerning the optimisation principle. After some background information on the foundation and evolution of the principle, this report describes the main characteristics of the process, addresses the issue of exposure distribution in that process, and provides the basic requirements for its application in operation and regulation. A description of decision-aiding techniques commonly used for practical implementation of the optimisation process is provided in Annex A.

Keywords: ALARA; BATNEEC; Collective dose; Safety culture; Stakeholder involvement

CONTENTS

PREFACE

On 20 October 2001, the Main Commission of the International Commission on Radiological Protection (ICRP) approved the formation of a new Task Group, reporting to Committee 4, to develop guidance on the principle and application of the optimisation of radiological protection. As stated in the terms of reference, the objective of the Task Group was to review the principle of optimisation and the requirements for its implementation in relation to the Commission's revised recommendations. In this perspective, particular attention had to be given to the role of constraints, the distribution of individual exposures, stakeholder involvement, and application in regulation and operation.

It was anticipated that the document produced as a result of the Task Group's work would form one of the foundation documents for the new recommendations. This report is the outcome of the Task Group's efforts, and it addresses the areas mentioned above. The guidance in this report builds upon the concept of optimisation of protection recommended previously by ICRP.

The membership of the Task Group was as follows:

W. Weiss (Chairman)	J.-F. Lecomte	Y. Xia
M.E. Clark	J. Lochard	

The corresponding member of this Task Group was:

T. Lazo

The Task Group would like to thank those organisations and staff that made facilities and support available for its meetings. These include the Federal Office for Radiation Protection in Germany, the French Institute for Radiation Protection and Nuclear Safety, and the Nuclear Energy Agency of the Organisation for Economic Co-operation and Development.

The report was adopted by the Commission at its meeting in Bern on 17 September 2005.

EXECUTIVE SUMMARY

(a) The optimisation of protection has been one of the fundamental principles of the system of radiological protection since the 1970s (ICRP, 1973, 1977). While the definition of this principle has remained relatively unchanged over time, its application has evolved with feedback of its practical implementation. Initially focused on quantitative techniques, mainly cost–benefit comparisons of protective options, the optimisation process incorporated progressively operational procedures, good practices, and qualitative approaches to become a more judgmental decision-making process.

(b) The principle of optimisation is defined by the Commission as the source-related process to keep the magnitude of individual doses, the number of people exposed, and the likelihood of potential exposure as low as reasonably achievable below the appropriate dose constraints, with economic and social factors being taken into account. According to the Commission's revised recommendations, this process of optimisation below constraint should be applied whatever the exposure situation; i.e. planned, emergency, or existing.

(c) The Commission's recommendations specifically related to the optimisation principle (ICRP, 1983, 1988), as well as the provisions about this principle given in *Publication 60* (ICRP, 1991), remain valid. Optimisation must be implemented through an ongoing, cyclical process that involves evaluation of the exposure situation to identify the need for action (the framing of the process), identification of the possible protective options to keep the exposure as low as reasonably achievable, selection of the best option under the prevailing circumstances, implementation of the selected option through an effective optimisation programme, and regular review of the exposure situation to evaluate if the prevailing circumstances call for the implementation of corrective protective actions. However, the way it should be implemented is now viewed more broadly to reflect the increasing role of individual equity, safety culture, and stakeholder involvement in our modern societies (ICRP, 1998, 1999).

(d) The optimisation principle as presented in this report is a consolidation and an evolution, but not a fundamental change, of the Commission's recommendations concerning this principle. The report addresses all exposure situations where radiological exposures are amenable to control, except patient exposures which are dealt with separately.

Characteristics of the process

(e) The optimisation of protection is a forward-looking iterative process aimed at preventing exposures before they occur. It is continuous, taking into account both technical and socio-economic developments, and requires both qualitative and quantitative judgments. The process should be systematic and carefully structured to ensure that all relevant aspects are taken into account. Optimisation is a frame of mind, always questioning whether the best has been done in the prevailing circumstances. It also requires commitment at all levels in all concerned organisations, as well as adequate procedures and resources.

(f) The process for assessing protective options, and for judging that no further dose reduction is reasonable, should involve the comparison of a number of feasible protective options to reduce the planned or potential doses to individuals and groups. A graded approach is necessary to take into account both the level of exposure and the complexity involved. Due to its judgmental nature, there is a strong need for transparency of the optimisation process. This transparency assumes that all relevant information is provided to the involved parties, and that the traceability of the decision-making process is properly documented, aiming for an informed decision.

(g) For the control of radioactive emissions to the environment, the principle of the best-available technology, not entailing excessive costs (BATNEEC), may be used. The principles of optimisation and BATNEEC complement each other. With a view to the consequences to human health, the control of residual exposures will be driven by the optimisation of estimated radiation doses.

(h) Procedures are necessary to clarify responsibilities for implementation of the optimisation process. At the operational level, an organisational structure should be established to organise a dialogue between the professional disciplines involved in an operation, including co-ordinators, working groups, or committees, regardless of whether or not the resulting structure is dedicated solely to optimisation.

(i) Commitment from all relevant parties, ranging from authorities to exposed individuals, to allow for effective implementation of optimisation implies:

- putting optimisation into regulation, willingness to enforce it, and providing guidelines with proper balance between dialogue and control (authorities);
- defining a radiological policy, setting general goals, developing and adhering to procedures, delegating responsibilities, allocating means and resources, and maintaining independence of radiological protection professionals from operation (operating management); and
- sharing information, maintaining vigilant attitude, training and retraining, and raising consciousness of radiological protection (individuals).

(j) The involvement of stakeholders (i.e. parties who have interests in and concern about a situation) is seen as an important input to the optimisation process. It is a proven means to achieve incorporation of values into the decision-making process, improvement of the substantive quality of decisions, resolution of conflicts among competing interests, building of shared understanding with both workers and the public, and building of trust in institutions. Furthermore, involving all concerned parties reinforces the safety culture, and introduces the necessary flexibility in the management of the radiological risk that is necessary to achieve more effective and sustainable decisions.

(k) The decision maker (generally the operating management or a competent authority) has clearly defined roles and responsibilities in the optimisation process. Other individuals and groups can also be considered as stakeholders. Examples include institutional and non-institutional technical support to the decision-making process (approved dosimetric services, qualified experts, formal technical services, public expert organisations, private laboratories), exposed individuals (either workers or members of the public) or their representatives (trade unions, local associations), and representatives of the society, either by an elective process (elected representatives) or a participative process (environmental associations). The involvement of stakeholders does not imply that the operating management and/or authorities relinquish their responsibility to make the final decision, or their accountability for that decision.

(l) The best option is always specific to the exposure situation and represents the best level of protection that can be achieved given the circumstances. Therefore, it is not relevant to determine, a priori, a dose level below which the optimisation process should stop. Depending on the exposure situation, the best option could be close to or well below the appropriate constraint. This means that the optimisation process may result in doses lower than any level that could be proposed as an 'entry level' into the system of radiological protection.

(m) It should be stressed that optimisation is not minimisation. It is the result of an evaluation that carefully balances the detriment from the exposure (economic, human, social, political, etc.) and the resources available for the protection of individuals. Thus, the best option is not necessarily the one with the lowest dose.

Optimisation and exposure distribution

(n) Comparison of protective options is a key feature of the optimisation process, which must entail careful consideration of the characteristics of the individual exposure distribution within a group of exposed population. Each group of population affected by a source can be described by different attributes, such as age, gender, and habits, as well as by various exposure parameters, such as mean, minimum, and maximum individual doses, the number of individuals exposed, the collective dose, and the likelihood of potential exposure. A single exposure parameter, however, is generally insufficient to fully compare the various protective options.

(o) Additional aspects to be considered in the comparison of protective options are social values, particularly equity in the distribution of exposure among the concerned group of individuals. For example, different protective options for a group of workers may be characterised by similar average individual and collective doses but rather different profiles of the dose distribution. In such a comparison, equity considerations will, in most cases, lead to discarding the protective options with the highest individual exposures.

(p) When the exposures occur over large populations, large geographical areas, and long periods of time, the total collective effective dose (i.e. the summation of all individual exposures in time and space) is not a useful tool for decision aiding because it may aggregate information excessively and could be misleading for selecting protective actions. To overcome the limitations associated with collective dose, each relevant exposure situation must be analysed carefully to identify the individual characteristics and exposure parameters that best describe the exposure distribution among the concerned population for the particular circumstance. Such an analysis results in the identification of various population groups with homogeneous characteristics for which collective doses can be calculated within the optimisation process.

(q) For identifying the relevant population groups, the same approach as that used for framing the optimisation process can be used. This includes asking when, where, and by whom exposures are received. The result of such questioning may be presented in a multidimensional collective dose matrix. Once this matrix is established, the relative importance of each element of the matrix, expressed as collective dose, can be weighted to reflect the economic and social considerations and values, as well as the preferences of those involved in the optimisation process.

The application of optimisation in operation and regulation

(r) Within the system of radiological protection, both the operators and the appropriate national authorities have responsibilities for applying the optimisation principle. Implementation of the process of optimisation of protection is the responsibility of the operating management, subject to the requirements of the authority. The operating management makes decisions regarding the design, organisation, and ongoing implementation of the optimisation process. The authority promotes and may require optimisation as a way to reach the level at which licence to operate, if any, can be granted. It may also verify that optimisation of radiological protection is implemented effectively during operation. The burden of proof of this implementation rests with the operating management. The decision to authorise an exposure-causing activity, or the implementation of exposure-reducing measures and their implied residual doses, rests with the authority. An active safety culture supports the successful application of optimisation by both the operating management and the authority.

(s) All aspects of optimisation cannot be codified; optimisation is more an obligation of means than of results. Except in cases of regulatory violation, it is not the role of the authority to focus on specific outcomes for a particular situation, but rather on processes, procedures, and judgments. A strong dialogue must be established between the authority and the operating management. The regulation should provide guidelines designed to build such a dialogue. The success of the optimisation process will depend strongly on the quality of this dialogue.

1. INTRODUCTION

(1) A basic requirement of the system of radiological protection recommended by the Commission is to optimise the level of protection achieved below source-related dose constraints (optimisation below constraint) whatever the exposure situation; i.e. normal, emergency, or existing.

(2) Historically, the optimisation of protection has been one of the fundamental principles of the system of radiological protection since the 1970s (ICRP, 1973, 1977). While the definition of this principle has remained relatively unchanged over time, its application has evolved with feedback of its practical implementation. Having started with its focus on quantitative techniques, mainly cost–benefit comparisons of protective options, the optimisation process progressively incorporated operational procedure, good practices, and qualitative approaches to become a more judgmental decision-making process.

(3) The Commission has published two reports that include recommendations specifically related to the optimisation principle (ICRP, 1983, 1988). These reports describe how optimisation is applied in various circumstances for the protection of workers and the public. In general, information in these reports is still relevant, particularly as it applies to quantitative methods for performing analyses.

(4) The way in which the optimisation principle is presented in this report is a consolidation and an evolution, but not a fundamental change, in the Commission's recommendations concerning this principle. The basic definition given in *Publication 60* (ICRP, 1991) remains valid, but the way in which it should be implemented is now viewed as a broader process reflecting the increasing role of individual equity, safety culture, and stakeholder involvement in our modern societies (ICRP, 1991, 1998, 1999).

(5) This report will address all exposure situations, i.e. normal, emergency, and existing, where radiological exposures are amenable to control, except for patient exposures. Particular attention is given to the treatment of individual dose distributions, to the respective responsibilities of the operating management and competent authorities, and to opportunities for the involvement of stakeholders in implementation of the optimisation principle. Furthermore, this report clarifies the respective roles of decision framing, decision aiding, and decision making in implementation of the optimisation process.

(6) Section 2 provides background information on the foundation and evolution of the optimisation principle. The characteristics of the optimisation process are described in Section 3. Section 4 addresses the role of exposure distribution in the optimisation process. Finally, Section 5 provides information about the application of optimisation in operation and regulation. The document is complemented with an annex on decision-aiding techniques for implementation of the optimisation process.

2. THE HISTORY OF THE OPTIMISATION PRINCIPLE

2.1. Foundation of the principle

(7) Introduction of the concept of optimisation in the recommendations of ICRP was a direct consequence of the recognition, during the 1940s, of the so-called 'stochastic effects', coupled with the impossibility of demonstrating the existence or non-existence of a threshold for this type of irreversible effects. Indeed, whilst the only recognised harmful effects of radiation were the deterministic effects, the limitation of exposure below the known thresholds for the appearance of these effects was considered to be sufficient to avoid any undesirable consequences of radiation. Due to the uncertainty of the dose–effect relationship for stochastic effects, the use of a limit was no longer a guarantee of the absence of risk. This led the Commission to adopt a prudent attitude and to recommend 'that every effort be made to reduce exposures to all types of ionising radiation to the lowest possible level' (ICRP, 1955, Para. VI). This position facilitated the Commission's introduction of the optimisation principle two decades later.

(8) The adoption of this prudent attitude for the management of stochastic effects raised the issue of justification of the exposure. In a context of uncertainty, imposing a risk on a group of individuals is only justified if there is a clear social benefit in return. Moreover, if an endeavour were to result in such a benefit, a second consideration is how far to reduce the risk and at the same time preserve the viability of the risk-causing activity. These considerations led the Commission to reword its first formulation and to recommend that 'all doses be kept as low as practicable and that any unnecessary exposure be avoided' (ICRP, 1959, Para. 45).

2.2. Evolution of the concept

(9) The next development in the optimisation principle was the elaboration of criteria for determining the level of exposure that can be considered 'as low as practicable'. Introduced in *Publication 9*, these were described in a new formulation of the earlier recommendation: 'As any exposure may involve some degree of risk, the Commission recommends that any unnecessary exposure be avoided, and that all doses be kept as low as is readily achievable, economic and social considerations being taken into account' (ICRP, 1966b, Para. 52). It was also stated in *Publication 9* that risk has two dimensions, individual and societal, and must be balanced with the benefits of the proposed activities. Furthermore, the objective of keeping exposure 'as low as readily achievable' must be balanced with the effort required to achieve this objective.

(10) Another key step in the principle's evolution was *Publication 22* (ICRP, 1973), which was devoted entirely to clarification of the statement above from *Publication 9*. In particular, the Commission introduced a cost–benefit model to help implement the principle in practice. The key point in *Publication 22* was the statement that 'It is possible to define the point at which it can be said that a dose is as low as is readily achievable, economic and social considerations taken into account, by choosing the

dose at which the economic and social gains of further reducing the dose are equal to the economic and social costs of achieving that reduction' (ICRP, 1973, Para. 11).

(11) Furthermore, the adverb 'readily' was replaced by 'reasonably' (ICRP, 1973, Para. 20) to describe more accurately the Commission's intent concerning the effort to be devoted towards risk reduction. Such an approach was made possible due to the availability of the first estimates of the magnitude of somatic and genetic risks associated with exposure to low doses and low dose rates published by the Commission in 1964 (ICRP, 1966a). The derived risk values per unit of exposure allowed the development of the concept of detriment, defined as the mathematical expression of 'the expectation of the harm incurred from a radiation dose' (ICRP, 1973, Para. 21). This concept constitutes one of the basic elements of the cost–benefit model for deciding whether or not a reduction in dose is reasonable.

(12) A minor change in the formulation was introduced in *Publication 26* (ICRP, 1977), where the term 'considerations' was replaced by 'factors'. Table 2.1 summarises the evolution of the wording for 'As Low As Reasonably Achievable – ALARA' over the past decades.

(13) For more than a decade, the cost–benefit model presented in *Publication 22* was the underlying concept of all methodological and practical developments for incorporating optimisation into the management of public and occupational exposures. The next significant step was *Publication 37* (ICRP, 1983), which was devoted to a mathematical presentation of the cost–benefit model as well as its practical use in the design and operation of installations.

(14) It soon became evident to those involved in the practical implementation of optimisation that decision making was driven by more parameters than those embodied in the strict cost–benefit approach. A first attempt to incorporate additional factors was the exploration of less rigid decision-aiding techniques, particularly those based on the scoring and ranking of multiple factors. A second approach was the development of procedures to assist operators with committing to ALARA.

(15) Both of these efforts were reflected in *Publication 55*, adopted by the Commission in 1988. Although this publication continued to apply theoretical developments

Table 2.1. Evolution of the optimisation principle

To **reduce** exposures	**to the lowest**	**possible** level		(ICRP, 1955)
To **keep** exposures	**as low as**	**practicable**		(*Publication 1;* ICRP, 1959)
To keep exposures	as low as	**readily achievable,**	**economic and social considerations** being taken into account	(*Publication 9;* ICRP, 1966b)
To keep exposures	as low as	**reasonably achievable,**	economic and social considerations being taken into account	(*Publication 22;* ICRP, 1973)
To keep exposures	as low as	reasonably achievable,	economic and social **factors** being taken into account	(*Publication 26;* ICRP, 1977)

and mathematical formulations, it also began the evolution towards a broader perspective for the decision-making process related to radiological protection and a more practical approach. For example, it stated that 'The concept of optimisation of protection is practical in nature. Optimisation provides a basic framework of thinking that is proper to carry out some kind of balancing of the resources put into protection, and the level of protection obtained, against a background of other factors and constraints, so as to obtain the best that can be achieved in the circumstances' (ICRP, 1988, Para. 8).

(16) Further evolution of the concept is present in the 1990 Recommendations adopted only 2 years later. In this publication, the Commission highlighted the need to consider 'the magnitude of individual doses, the number of people exposed and the likelihood of incurring exposures where these are not certain to be received' when implementing the optimisation process. Moreover, the emphasis was placed on the equity issue raised by the uneven distribution of benefits and detriments through society. In this perspective, it is recognised that the 'optimisation of protection may thus introduce a substantial inequity between one individual and another' (ICRP, 1991, Para. 121). The Commission addressed these considerations by introducing the concept of dose constraint as 'the source-related values of individual dose used to limit the range of options considered in the procedure of optimisation', but recommended its use to practices only and not to intervention (ICRP, 1991, Para. 144). It is also noticeable in this publication that, beyond the strict cost–benefit model, the Commission insisted on the importance of informal processes and practical procedures to keep exposure as low as reasonably achievable.

2.3. Recent developments

(17) Several publications since *Publication 60* have introduced new elements concerning optimisation in relation to its application in various contexts. For example, *Publication 63*, on the principles for intervention for protection of the public in a radiological emergency (ICRP, 1993), emphasised the key role of optimisation in the design of protective actions for mitigating the consequences of accidents. The optimisation principle is also the major focus of *Publication 75*, devoted to the protection of workers (ICRP, 1998). The developments in this publication stress the importance of managerial arrangements in the practical implementation of optimisation for protection at work and, particularly, an explicit commitment to a safety-based attitude. In *Publication 77* (ICRP, 1997), which addresses the radiological protection policy for the disposal of radioactive waste, the Commission reiterates the judgmental nature of the optimisation principle and emphasises the possible misuses of the collective dose concept for the purpose of comparing protective options dealing with small individual doses spread over a very long time. In *Publication 81* (ICRP, 1998), dealing with the disposal of long-lived solid radioactive waste, the Commission recommends going beyond the quantitative approaches developed during the 1970s and 1980s, and advocates adopting a broader perspective.

(18) Another important move in this direction is taken by the Commission in *Publication 82* (ICRP, 1999), on the protection of the public in situations of prolonged radiation exposure. In this publication, the Commission reiterates that it provides recommendations on radiological protection on the basis of objective assessments of the health risks associated with exposure levels and relevant attributes of various exposure situations. It also recognises, however, the reality of sociopolitical and cultural considerations that usually influence the final decision on the level of protection. As a consequence, the Commission anticipates that the decision-making process 'may take into account attributes other than those directly related to radiological protection' and 'will include the participation of relevant stakeholders rather than radiological protection specialists only' (ICRP, 1999, Para. 4).

(19) Following these recommendations, analyses of practical experiences at national and international level have allowed a better understanding of the challenges, implications, and benefits associated with greater stakeholder involvement in radiation protection decision-making processes (NEA, 1998, 2001, 2004). As a result, the Commission now considers that the involvement of stakeholders is an important input of the optimisation process, because it introduces the necessary flexibility in the management of the radiological risk that is necessary to achieve more effective and sustainable decisions.

3. THE OPTIMISATION PROCESS

(20) The principle of optimisation of radiological protection is defined by the Commission as the source-related process to keep the magnitude of individual doses, the number of people exposed, and the likelihood of potential exposure as low as reasonably achievable below the appropriate dose constraints, with economic and social factors being taken into account.

(21) It is not possible to give a simple formal definition of a single source or of the total group of relevant sources. In the application of optimisation below constraint, the term 'single source' should be used in a broad sense, such as the x-ray equipment in a hospital, or the release of radioactive materials from an installation. Most situations will give rise to a predominant source of exposure for any single individual, or representative person, making it possible to treat sources separately when considering actions. Provided that the operating management and the regulators both apply the Commission's broad policies, the definition of a single source is straightforward. Difficulties will arise if the policy is distorted, e.g. by artificially subdividing a source in order to avoid the need for protective action, or by excessively aggregating sources to exaggerate the need for action.

(22) To provide the best protection under the prevailing circumstances (in normal, emergency or existing controllable situations), this process of optimisation below constraint must be implemented through an ongoing, cyclical process (called the optimisation process) that involves evaluation of the exposure situation to identify the need for action (framing of the process); identification of the possible protective options to keep the exposure as low as reasonably achievable; selection of the best option under the prevailing circumstances; implementation of the selected option through an effective optimisation programme; and regular review of the exposure situation to evaluate if the prevailing circumstances call for the implementation of corrective protective actions.

(23) Implementation of the optimisation principle of protection is a process that is at the heart of a successful radiological protection programme. It must be framed carefully to take into account the relevant attributes of the exposure situation. Furthermore, it should include, as appropriate to the exposure situation, the involvement of the relevant stakeholders. These two features, which were introduced in *Publication 82* (ICRP, 1999), are considered by the Commission as important components of the optimisation process.

3.1. Framing the process

(24) The objective is to identify clearly and systematically the relevant attributes necessary to select the best protective options under the circumstances. In this respect, the characteristics of the exposure distribution (i.e. individual doses, mean dose, number of people exposed) are only one part of the attributes to be considered.

(25) To identify the relevant attributes, the most straightforward approach is to ask 'when, where, how, and by whom are exposures received'. Responding to these questions will result in a set of attributes expressing the characteristics of exposed

populations and their exposures, as well as technical, economic, social, environmental, and ethical considerations relevant to the situation. The Commission recommends that attention should be given to the avoidance of accidents or any potential exposure, transfer of exposure between groups, and the distribution of exposures over long time periods and distant populations. For many situations, the participation of stakeholders in the framing process is an aid to help identify relevant attributes.

(26) A representative list of useful attributes to consider in selection of the best protective option is presented in Table 3.1. This list is not exhaustive and other aspects may need to be included depending upon the specific context of the exposure situation. On the other hand, although the list is not exhaustive, it may include too many attributes relevant to a given situation. In many situations, a limited number of attributes will be sufficient. Therefore, case-specific selection of relevant attributes is required for each situation to address the key options properly. Consideration of a wider spectrum of attributes related to the exposure situation is important for comprehensive evaluation of the situation.

3.2. Characteristics of the process

(27) The optimisation of protection is a forward-looking iterative process aimed at preventing exposures before they occur. It is continuous, taking into account both technical and socio-economic developments, and requires both qualitative and quantitative judgments. The process should be systematic and carefully structured to ensure that all relevant aspects are taken into account. Optimisation is a frame of mind, always questioning whether the best has been done in the prevailing circumstances. It also requires commitment at all levels in all concerned organisations as well as adequate procedures and resources.

(28) The process for assessing protective options, and for judging that no further dose reduction is reasonable, should involve the comparison of a number of feasible protective options to reduce the planned or potential doses to individuals and groups. Measures taken to protect individuals or groups from a source of radiation can be applied at the source, in the environment between the source and the individual, or to the individual. Where feasible, controls applied at the source are preferable. Such measures are less disruptive and they apply to all pathways for all individuals for any source. In contrast, controls applied to the environment or to individuals may not be all inclusive. Furthermore, at least regarding public exposure, there is less likelihood of unexpected socio-economic implications to measures that are source related.

(29) The optimisation of protection is a systematic process that needs to take a long-term view with regard to 'cradle to grave considerations', e.g. optimisation during the design phase of an installation must also consider all the following phases of operation of the facility, including decommissioning. For normal and existing situations, much of the protection is built-in during the design phase of a project for controllable sources, when options are evaluated, often for the selection of engineered controls. The process of optimisation of protection must continue during the operational and termination phases. In emergency situations, optimisation should be

Table 3.1. Representative attributes to select the best protective option (non-exhaustive list)

- Characteristics of the exposed population
  Gender
  Age
  Health status
  Sensitive groups (e.g. pregnant women)
  Habits

- Characteristics of the exposure
  Distribution of exposures in time and space
  Number of individuals
  Minimum individual dose
  Maximum individual dose
  Mean individual dose
  Statistical deviations
  Collective dose associated with ranges of individual doses
  Likelihood of potential exposure
  Pre-existing radiological conditions (e.g. high natural background, enhanced level of exposure due to past activities or accidents)

- Social considerations and values
  Equity
  Ability to control (measurements, health surveillance, etc.)
  Sustainability
  Intergenerational consideration
  Individual benefit
  Social benefit
  Level of information/knowledge held by those exposed
  Social trust

- Environmental considerations
  Impact on fauna and flora
  Impacts on climate

- Non-radiation hazards

- Technical and economic considerations for protective options
  Feasibility
  Costs
  Uncertainties

- Political aspects

- Regulatory constraints

used at the planning phase to identify protective options and to select appropriate levels of constraints. During an emergency, the process of optimisation should be applied in a flexible manner to take the actual circumstances into account. In existing controllable situations, optimisation is generally implemented through a step-by-step approach that can last over long periods of time (e.g. post-accident situation, radon reduction programmes). The optimisation process incorporates a wide range of qualitative and quantitative methods and tools. Several of them, such as measurements, models, checklists, real-time software, on-the-job analyses, operational dosimetry systems, radiological performance targets, documentation, databases, decision-aiding tools, and the reference monetary value of the man sievert, are commonly used and have been presented in *Publications 37* and *55* (ICRP, 1983, 1988). Quantitative methods may provide valuable input to the decision-making process. However, given the many qualitative factors, they should never be the sole input (see Annex A). Due to uncertainties, approximations, pragmatic concerns, technical and economic restrictions, or conflicting societal values, a qualitative judgmental approach is also necessary. In many situations, such an approach may usefully complement approaches based on decision-aiding techniques relying on quantitative data. In the decision-aiding process in particular, the involvement of the relevant stakeholder is becoming increasingly recognised as an effective input (see Section 3.3).

(30) The optimisation process should be as elaborate as necessary to address a given situation. A graded approach is needed to take into account both the level of exposure and the complexity involved. For many exposure situations, decisions can be made easily using sound methods, tools, and professional skills. However, past experience shows that complex and long-lasting processes are sometimes necessary to arrive at protection decisions related to relatively low levels of exposures when economic, social, and political considerations are dominant.

(31) During an operation, the process of optimisation of protection is also ongoing, and questions are raised regarding whether enough has been done before, during, and after the exposure. As shown in Fig. 3.1, the optimisation process is cyclical. It is essential that reviews are planned and implemented at regular time intervals. Past performance, trend analyses of dose (or other data), results of internal audits, peer reviews, incident reports, and lessons learned all feed into this process. When the selected option is implemented, reviews may indicate that the results differ from those expected. In such circumstances, a new evaluation cycle may be necessary. The methods used to judge if protective options are reasonable may also change with time.

(32) There is a need for any optimisation process to make numerous decisions related to radiological protection, taking into account a number of attributes, such as technical feasibility, cost, social factors, potential adverse impacts, long-term effectiveness, and the public's or workers' concerns, as well as their relative importance. Such decisions include whether an action is really necessary, which option is the most effective and efficient, and what resources are reasonable to complete the undertaking.

(33) Optimisation is a frame of mind. The effective implementation of the process entails that all stakeholders involved know and agree with the basic assumptions of radiological protection. The acknowledgement that any level of exposure can induce a risk should be the incentive to ensure that all those involved in the optimisation process are accountable for its effective implementation. Furthermore, they should adhere to an active safety culture, the key attributes of which are 'personal dedication, safety thinking and an inherently questioning attitude (...) Good practices in themselves, while an essential component of safety culture, are not sufficient if applied mechanically. There is a requirement to go beyond the strict implementation of good practices so that all duties important to safety are carried out correctly, with

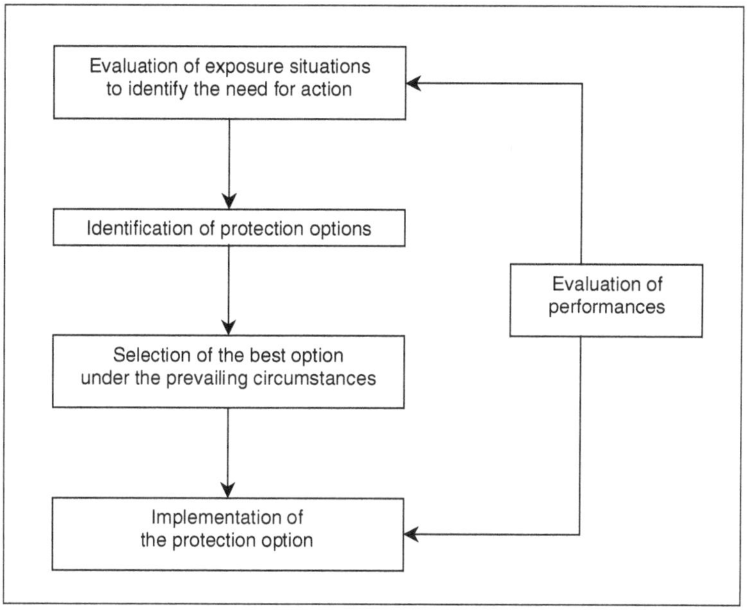

Fig. 3.1. Schematic view of the optimisation process.

alertness, due thought and full knowledge, sound judgment and proper sense of accountability' (IAEA, 1991).

(34) Due to its judgmental nature, there is a strong need for transparency of the optimisation process. All the data, parameters, assumptions, and values that enter into the process must be presented and defined very clearly. This transparency assumes that all relevant information is provided to the involved parties, and that the traceability of the decision-making process is documented properly, aiming for an informed decision.

(35) For the control of radioactive emissions to the environment, the BATNEEC principle may be used. The principles of optimisation and BATNEEC complement each other. With a view to the consequences to human health, the control of residual exposures will be driven by optimisation of estimated radiation doses. With regard to the control of effluent releases, or in situations where humans are not affected directly or are not the primary protection target, the optimisation will generally apply BATNEEC to control effluent releases. Since the Rio Conference (UN, 1992), the central organising principle of international environmental policy, sustainable development, has moved beyond health-driven emission standards towards BATNEEC techniques in the non-radiological sector, focusing on reducing or, where practical, eliminating emissions at the source. This approach is being applied increasingly to facilities and operations, focusing the objective of protection on emission reduction rather than on health-effect/probability-of-effect reduction. The BATNEEC concept with due consideration to social and economic factors is

close to the Commission's recommendation to keep doses as low as reasonably achievable (ICRP, 1997, 1998).

(36) Procedures are necessary to clarify responsibilities for the implementation of the optimisation process. At the operational level, an organisational structure should be established to organise a dialogue between the professional disciplines involved in an operation, including co-ordinators, working groups, or committees, whether or not the resulting structure is dedicated solely to optimisation.

(37) Finally, the effective implementation of optimisation requires commitment from all relevant parties, ranging from authorities to exposed individuals. Elements required to ensure this commitment include:

- putting optimisation into regulation, willingness to enforce it, and providing guidelines with proper balance between dialogue and control (authorities);
- defining a radiological policy, setting general goals, developing and adhering to procedures, delegating responsibilities, allocating means and resources, and maintaining independence of radiological protection professionals from operation (operating management); and
- sharing information, maintaining vigilant attitude, training and retraining, and consciousness raising in radiological protection (individuals).

The respective responsibilities in implementing these provisions are presented in more detail in Section 6 of (IAEA, 2002).

3.3. Stakeholder involvement

(38) The involvement of stakeholders, a term which was introduced by the Commission in *Publication 82* (ICRP, 1999) to mean those parties who have interests in and concern about a situation, is seen as an important input to the optimisation process. The decision maker (generally the operating management or a competent authority) has clearly defined roles and responsibilities in this process. Other individuals and groups can also be considered as stakeholders. Examples include the exposed individuals (either workers or members of the public) or their representatives (trade unions, local associations), institutional and non-institutional technical support to the decision-making process (approved dosimetric services, qualified experts, formal technical services, public expert organisations, private laboratories), and representatives of the society, either by an elective process (elected representatives) or a participative process (environmental associations).

(39) The involvement of stakeholders is a proven means to achieve incorporation of values into the decision-making process, improvement of the substantive quality of decisions, resolution of conflicts among competing interests, building of shared understanding with both workers and the public, and building of trust in institutions. Furthermore, involving all concerned parties reinforces the safety culture and introduces the necessary flexibility in the management of the radiological risk that is needed to achieve more effective and sustainable decisions. Stakeholders may be particularly helpful for identification of the attributes of the exposure situation and their

relative importance, as well as for identification of protective options within the framing of the decision-making process.

(40) The extent of stakeholder involvement will vary from one situation to another. Depending upon the circumstances, it may not be necessary to involve all stakeholders, or types of stakeholders, in every aspect or phase of the optimisation process. Many radiological protection decisions will not be complex or socially contentious, and thus will not need broad stakeholder involvement. While there is no unique approach for developing stakeholder involvement, experience is increasing. Various methods have been developed in different areas to structure the process of linking stakeholders to the decision-making process. The spectrum covers classical consultation processes at one end and structured consensus building techniques with or without assistance of a third party at the other end (Beierle, 2002; NEA, 2004).

(41) The involvement of stakeholders does not imply that the operating management and/or authorities relinquish their responsibility to make the final decision, or their accountability for that decision. The question of final responsibility for decisions must not be obscured during the shared steps of decision framing and implementation of the optimisation process. Responsibility for the 'final decision' with respect to the adequacy of protection solutions ultimately lies with the operating management and/or the authority.

3.4. Selection of the best option

(42) The best option is always specific to the exposure situation and represents the best level of protection that can be achieved given the circumstances. Therefore, it is not relevant to determine, a priori, a dose level below which the optimisation process should stop. Depending on the exposure situation, the best option could be close to or well below the appropriate constraint. This means that the optimisation process may result in doses lower than any level that could be proposed as an 'entry level' into the system of radiological protection.

(43) In some cases, the technical, economic, legal, or social contexts may change optimisation solutions that have been agreed previously. For example, introduction of new technologies, increasing public concern, or availability of new resources for protection will be incentive for revisiting the situation, implementing new protective options, and possibly setting a new endpoint. Such changes should be addressed on a case-by-case basis, as has been done in the area of construction codes and fire protection regulations for buildings.

(44) Finally, it should be stressed that optimisation is not minimisation. It is the result of an evaluation that carefully balances the detriment from the exposure (economic, human, social, political, etc.) and the resources available for the protection of individuals. Thus, the best option is not necessarily the option with the lowest dose.

4. EXPOSURE DISTRIBUTION

(45) The comparison of protective options is a key feature of the optimisation process that must entail careful consideration of the characteristics of the individual exposure distribution within a group of exposed population. Each group of the population affected by a source can be described by different attributes, such as age, gender, and habits, as well as by various exposure parameters, such as mean, minimum, and maximum individual doses, the number of individuals exposed, the collective dose, and the likelihood of potential exposure. A single exposure parameter, however, is generally insufficient to compare the various protective options fully.

(46) Additional aspects to be considered in the comparison of protective options are the social values, particularly equity in the distribution of exposure among the concerned group of individuals. For example, different protective options for a group of workers may be characterised by similar average individual and collective doses but rather different profiles of the dose distribution. In such a comparison, equity considerations will, in most cases, lead to protective options with the highest individual exposures being discarded.

(47) For occupational exposure situations, information about individual doses to workers is accessible in most cases, and assessment of the individual dose distribution is relatively easy. For public exposure situations, information about individual doses is generally not directly accessible and these can only be estimated using surrogates. For example, modelled average individual doses can be estimated for different subgroups exposed to a given source. For such an approach, it is necessary to define the place inhabited (distance from the source), age and gender distribution, and living habits (diet, types of recreation) for each group of exposed individuals. If necessary, it is also possible to estimate the evolution of exposure in time for each group for the current and future generations.

4.1. Use of collective dose

(48) One way to characterise the distribution of individual exposures within groups for the purpose of comparing protective options in the optimisation process is by using the collective dose associated with this distribution. This concept was developed in response to the need to take into account the global impact of a given source on the population. Historically, it emerged from concerns related to the fallout from nuclear weapon tests and radioactive releases into the environment associated with the development of the nuclear power industry. It was introduced in the 1970s to serve as a basis for restricting the uncontrolled build-up of exposure to long-lived radionuclides in the environment, and for facilitating implementation of the cost–benefit analysis proposed at the time to implement the optimisation principle (IPSN, 2002).

(49) The collective dose is a measure of radiation exposure from a source in a given group of population. It is the integral of the distribution of the individuals' doses within this group. In *Publication 60*, the Commission recommends that one should

take account of the number of people exposed by multiplying the average dose to the exposed group by the number of individuals in the group (ICRP, 1991, Para. 34). If several groups of population are involved, the total collective effective dose associated with a source or an exposure situation is defined as the summation of the collective effective doses in all groups exposed by this source or this situation.

(50) In the case of occupational exposure, the collective dose is commonly used as a 'performance indicator' to characterise the total exposure associated with the operation of installations over a given period of time or with a particular type of work. For the purpose of comparison of protective options in the optimisation process, the collective dose is not always sufficient to characterise the individual dose distribution, especially when significant differences exist in the magnitude of individual exposures within the exposed group. In such circumstances, equity considerations need to take into account both the individual and collective doses associated with the distribution of exposure (see Annex A).

(51) In the case of public exposure, the collective dose may be a useful input to the optimisation process when the individual dose distributions are relatively homogenous and well defined. However, depending on the source, the radiological impacts can be more or less spread geographically (from local impacts to regional and, in some circumstances, even larger areas) and in time (from short term to mid term or sometimes long term), and covering a large range of individual doses. While collective doses can be estimated based on pathway assumptions in such situations, the value of such collective dose estimations for protection decisions is somewhat limited (ICRP, 1997, 1998).

4.2. Exposure distributions in time and space

(52) When the exposures occur over large populations, large geographical areas, and large periods of time, the total collective effective dose, as defined above (i.e. the summation of all individual exposures in time and space), is not a useful tool for decision aiding because it may aggregate information excessively and could be misleading for selecting protective actions. A straightforward application of the collective dose concept can mask the exposure characteristics of the dose distribution, as well as the inherent uncertainties attached to the dose assessment. Furthermore, it does not allow for due consideration of important sociopolitical considerations, e.g. equity, that may be particularly important for evaluating and comparing options.

(53) To overcome the limitations associated with collective dose, each relevant exposure situation must be analysed carefully to identify the individual characteristics and exposure parameters that best describe the exposure distribution among the concerned population for the particular circumstance. Such an analysis results in the identification of various population groups with homogeneous characteristics to be considered within the optimisation process. In particular, a collective dose resulting from a very wide range of individual doses should be disaggregated into a series of collective doses corresponding to homogeneous parts of the dose distribution (ICRP, 1997, 1998). The appropriate characteristics for defining the population groups

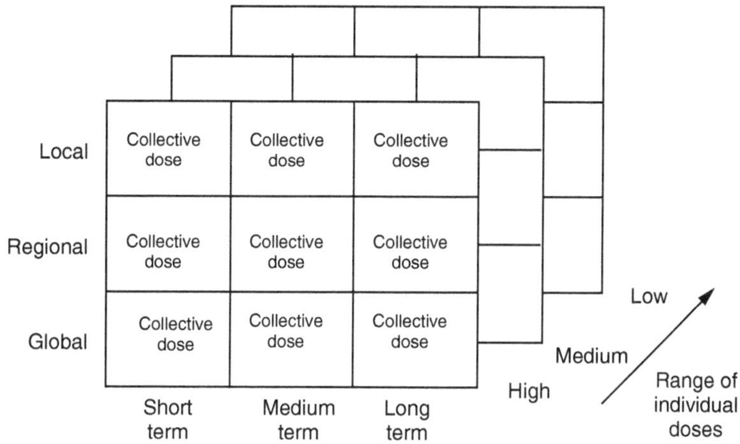

Fig. 4.1. Illustration of a collective dose matrix.

corresponding to these collective doses must be made on a case-by-case basis according to the exposure situation.

(54) The same approach as that used for framing the optimisation process (see Section 3.1) can be used for identifying the relevant population groups. This includes asking when, where, and by whom exposures are received. As an illustration, Fig. 4.1 shows the result of such questioning in the case of a dose distribution spread over time, space, and for various ranges of individual doses. The result is presented in a three-dimensional collective dose matrix:

- the space distribution of the exposed population is represented at various distances from the source: local, regional, or global;
- the time distribution of the exposed population is represented for various terms: short, medium, and long term. For exposures that are expected to last over very long periods, the time frame could also be expressed over a series of generations; and
- the distribution of individual doses of the exposed population is represented by various ranges in millisieverts: high (above 10), medium (between 10 and 1), and low doses (below 1).

Each element of the matrix corresponds to a collective dose associated with a given group. Other characteristics, such as age, gender, socioprofessional categories, or specific habits, could be used in this type of matrix if judged relevant for comparison of the protective options.

4.3. Collective dose matrix and decision-making process

(55) Once the collective dose matrix is established, the relative importance of each element of the matrix can be weighted to reflect the economic and social considerations and values, as well as the preferences of those involved in the optimisation

process. The nature of the considerations and values may vary significantly from case to case, as will the importance given by the stakeholders to each element. The degree of uncertainty in the level of exposure and any other relevant factors may also be considered. For example, to take equity considerations into account, relative weights can be assigned to each collective dose based on the magnitude of the mean individual dose that characterises a given group of population. This may be a way to give more importance to groups of individuals receiving higher doses than to groups of individuals receiving lower doses (see Annex A).

(56) Weights can also be assigned according to the time at which the exposure is predicted to occur. As there are uncertainties in the estimation of the dose and the associated detrimental increase for exposures to be received in the far future, the use of predicted exposures for decision-making purposes becomes increasingly problematic (ICRP, 1997). Consequently, progressively less importance could be given to individual exposures received in the far future due to the increasing uncertainties both in the estimation of the dose and in the associated detriment. The current relationship between dose and detriment may no longer be valid for future populations. Conversely, in particular exposure situations, more importance could be given to exposures occurring in the future based on intergenerational equity considerations or on not yet expected scientific evidence. Another judgment could be that exposures should be equally weighted in time. The Commission feels that our current state of knowledge and our ability to predict populations and exposure pathways can contribute appropriately to decision making for exposures occurring over a time period covering a few generations. Beyond such time frames, the Commission recommends that predicted doses should not play a major part in decision making.

5. THE APPLICATION OF OPTIMISATION IN OPERATION AND REGULATION

(57) Within the system of radiological protection, both the operators and the appropriate national authority have responsibilities for applying the optimisation principle. Implementation of the optimisation of protection is the responsibility of the operating management, subject to the requirements of the authority.

(58) This is the case for all controllable sources and exposure situations; i.e. planned, emergency, and existing. It should be noted, however, that the notions of 'operating management' and 'competent national authorities' should be interpreted broadly in these three situations, more along the lines of 'implementing organisation' and 'decision maker'.

(59) An active safety culture supports the successful application of optimisation, and both the operating management and the authority have essential roles to play in ensuring that an effective safety culture is developed and maintained. In particular, the authority should encourage the operating management to develop a 'safety culture' within their organisations. Such a safety culture should also exist within the authority.

(60) The operating management makes decisions regarding the design, organisation, and ongoing implementation of the optimisation process. The authority promotes and may require optimisation as a way to reach the level at which licence to operate, if any, can be granted. They may also verify that optimisation of radiological protection is implemented effectively during operation. The burden of proof of this implementation rests with the operating management. The decision to authorise an exposure-causing activity, or the implementation of exposure-reducing measures and their implied residual doses rests with the authority. In some cases, work is planned, assigned, performed, and overseen by others who are not under the direct control of the operating management. In such circumstances, any sharing of responsibility for optimisation should be documented clearly and understood fully by all parties (NEA, 1997).

(61) The operating management should develop and provide internal policies, priorities, rules, procedures, and quality assurance programmes to ensure the existence of a solid safety culture at all levels of management and staff. In this context, the objective of operating management is to prevent accidents, manage the probability of potential exposures, and keep worker and public exposures as low as reasonably achievable, with social and economic factors being taken into account.

(62) All aspects of optimisation cannot be codified; optimisation is more an obligation of means than of results. Except in cases of regulatory violation, it is not the role of the authority to focus on specific outcomes for a particular situation, but rather on processes, procedures, and judgments. A strong dialogue must be established between the authority and the operating management. The regulation should provide guidelines designed to build such a dialogue. The success of the optimisation process will depend strongly on the quality of this dialogue.

ANNEX A: OPTIMISATION AND DECISION-AIDING TECHNIQUES

A.1. Introduction

(A1) The use of decision-aiding techniques to quantify and compare protective options in implementation of the optimisation process is a well-established approach. It allows those who have to decide about the level of protection to select the best trade-offs between the various attributes involved in the process, taking the inherent uncertainties and value judgments into account. According to the degree of complexity of the situation to which the options may apply, different techniques can be applied.

(A2) Historically, cost–benefit analysis was the first technique promoted by the Commission in the early 1970s to balance the cost of the radiological detriment and the cost of protective measures (ICRP, 1973). It is a straightforward method that can be applied to many areas of public and occupational protection in normal, emergency, or existing situations. Later, as the optimisation principle was gaining support and its practical implementation was growing, other decision-aiding techniques were proposed by the Commission, such as cost-effectiveness or multi-attribute analysis (ICRP, 1989).

(A3) Fig. A1 presents a schematic overview of the successive steps of the optimisation process (cf. Section 3.2) and specifies where decision-aiding techniques fit in. It is important to note that the use of decision-aiding techniques is just an input into

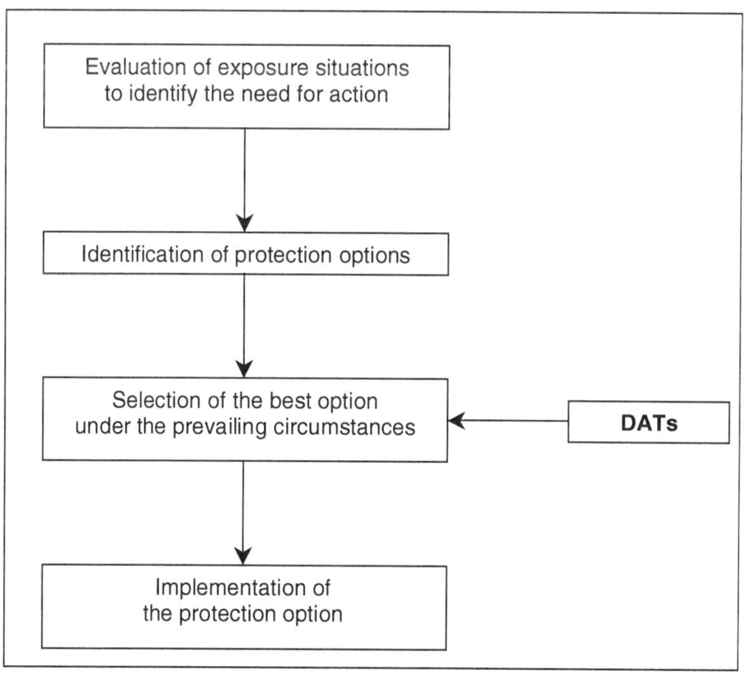

Fig. A1. The optimisation process and decision-aiding techniques (DATs).

the decision process. The important part of it is the initial framing that allows identification of all the relevant attributes to be taken into account, and identification of all possible protective actions to be evaluated. An important feature related to the use of decision-aiding techniques for selection of the best option under the prevailing circumstances is the need to express the various attributes in quantitative terms. This quantification process is, in many cases, the most difficult and time-consuming step, where all necessary data have to be gathered or produced through modelling and value judgments made concerning intangible attributes.

(A4) As far as the implementation of decision-aiding techniques is concerned, the choice of a particular technique depends mainly on the scope of the exposure situation, i.e. the diversity of the various attributes and judgments to be incorporated, but also the degree of their quantification and the importance of the uncertainties attached to the data characterising each protective option.

(A5) Moreover, whatever the technique that is finally selected, there is a need to introduce criteria for comparison and selection of the option. The most well-known criterion is the monetary value of the man sievert, which allows one to directly balance the economic cost of improving protection with the benefit in terms of dose reduction (ICRP, 1973). The following paragraphs give a brief description of the three basic decision-aiding techniques commonly used for practical implementation of the optimisation process, as well as the concept of the monetary value of the man sievert.

A.2. Cost–benefit analysis

(A6) There are different ways to perform a cost–benefit analysis (ICRP, 1983). The most straightforward technique is to express, in monetary terms, the various factors influencing the balance between the costs on one side and the benefits on the other side, and to aggregate them in order to select the option with the lowest monetary value of this aggregate. A key element in this procedure, when applied to the selection of radiological protective options, is the use of the monetary value of the man sievert, which allows expression of the benefit of protection (i.e. the reduction in dose related to the implementation of a protective option) in the same unit as the protective costs (see Section A.3).

(A7) A simple formulation of the cost–benefit analysis is to express the economic cost of the collective dose (Y) as follows:

$$Y = \sum \alpha_j S_j$$

where α_j is the monetary value of the man sievert to be applied to the group of population or workers j; S_j is the collective dose of the exposed group of population or workers j; and j may depend on the category of population exposed, the spread of exposure in time, and the level of individual dose (see Section 4).

(A8) The total cost of each option is calculated as the sum of its associated cost of protection (X) and the corresponding cost of the collective exposure (Y). The optimum protective option is given by the minimum value of the total cost ($X + Y$) as illustrated in Fig. A2. It is important to note that at the optimum level of protection,

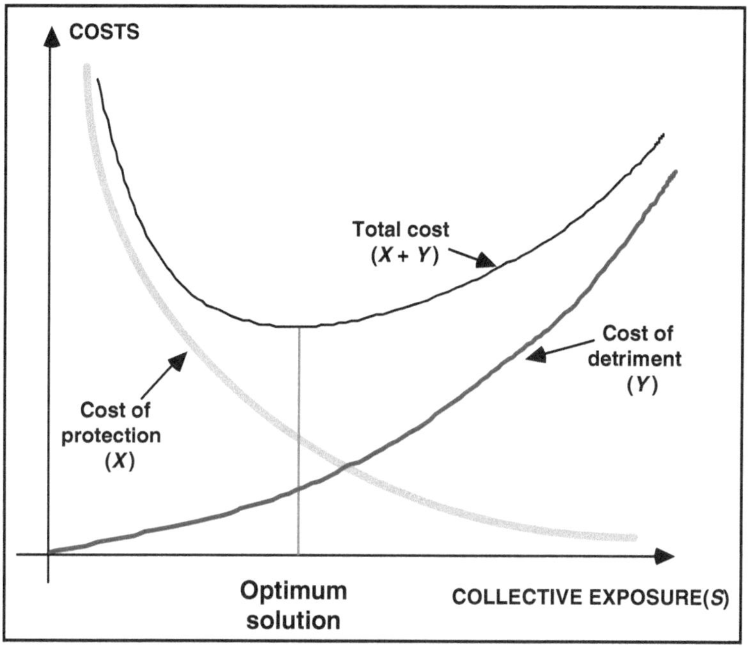

Fig. A2. Cost–benefit analysis.

the marginal cost of protection is equal to the marginal cost of the unit of collective dose avoided.

A.3. Monetary value of the man sievert

(A9) The definition and use of the monetary value of the man sievert has been, and remains, a matter of debate since the concept was introduced formally by ICRP in *Publication 22* (ICRP, 1973). First, there have always been some reservations about the use of the concept from an ethical point of view because of its link with the economic 'value of life'. Secondly, the methodological tenets on which it relies for its quantification have never found a broad consensus.

(A10) Despite these reservations, the concept was further developed with the practical implementation of the optimisation principle, and many organisations (operators and authorities) in the world have adopted values to be used more or less formally in the optimisation process in relation to decision-aiding techniques.

(A11) A major step in the development of the concept was the introduction of the concept of risk aversion in *Publication 37* (ICRP, 1983). With the risk aversion concept, allowance can be made for considerations concerning individual risk, i.e. individual doses within an exposed population. Thus, a collective dose of one man sievert resulting from ten individual doses of 100 mSv and the same collective dose resulting from 1000 doses of 1 mSv will not be assessed in the same way as regards monetary value, even though the potential health risk is identical from a collective point of view, given that the hypothesis of a linear no-threshold dose–effect relationship has been adopted. When the risk for individuals increases, there is a general tendency to be more protective and a consequence to be ready to allocate more resources and to reduce the risk.

(A12) Since *Publication 60* (ICRP, 1991), the Commission has placed increasing emphasis on equity considerations with regards to dose distributions, insisting on the fact that both the magnitude of individual doses and the number of people exposed should be kept as low as reasonably achievable below the relevant dose constraints, with economic and social factors being taken into account.

(A13) In order to introduce the considerations of risk aversion and equity in the distribution of individual dose levels in the monetary value of the man sievert, a model was developed in the 1990s (Lochard et al., 1996). The proposed model was designed in such a way as to reduce the collective exposure associated with a dose distribution of a given group of population or workers, while at the same time reducing the spread of the distribution and the highest individual doses of the dose distribution.

(A14) From an analytical point of view, the model is formalised as follows:

$$\alpha_{\text{ref}}(d) = \alpha_{\text{base}}\left(\frac{d}{d_0}\right)^a$$

where $\alpha_{\text{ref}}(d)$ is the monetary value of the man sievert for individual exposure level d; α_{base} is the basic monetary value of the man sievert; d_0 is the lower value of the individual dose range from which the aversion phenomenon can be applied; d is the annual individual exposure level; and a is the coefficient representing the degree of aversion ($a = 0$ when $d < d_0$, $a \geqslant 0$ when $d \geqslant d_0$).

(A15) It is therefore a system whereby the monetary values of the man sievert increase as individual exposure levels increase. The basic monetary value of the man sievert (α_{base}) reflects the value of the expected health effects. Regardless of the level of individual exposure, the monetary value of the health effects associated with one man sievert is considered constant. Coefficient 'a' represents the degree of risk aversion and makes it possible to introduce monetary values for the man sievert that increase as a function of individual exposure level. As for the lower limit d_0, it means that risk aversion can be considered only beyond a certain minimum exposure level, it being considered unnecessary to make allowance for individual exposure breakdown in the case of low dose ranges.

(A16) Fig. A3 presents the model. The y-axis shows what it is reasonable to spend to prevent one man sievert, expressed in monetary units, and the x-axis shows the individual dose levels in millisieverts.

(A17) In practice, in order to implement this model, it is necessary to define a value for the three parameters 'α_{base}', 'd_0' and 'a':

- The value of 'α_{base}' represents the monetary value of the health detriment associated with one unit of collective exposure, i.e. the loss of life expectancy associated with one man sievert. Various economic methods, including the Human Capital Approach (based on Gross National Product), can be used for this evaluation (Stokell et al., 1991).

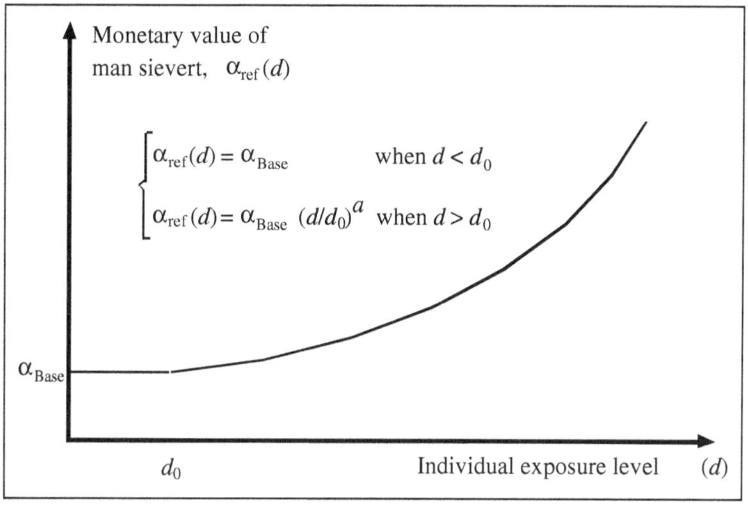

Fig. A3. Model of monetary values of the man sievert incorporating risk aversion and equity consideration.

- The value of 'd_0' corresponds with the upper level of individual dose below which aversion to the dispersion of exposure is not considered. This value depends upon the degree of acceptability of risk for the exposed population. In case of occupational exposure for example, it seems reasonable to adopt the value corresponding to the relevant primary constraint for public exposure in normal situations (1 mSv/year).
- The '*a*' coefficient reflects the degree of aversion to dispersion of individual exposure. It can be demonstrated that '*a*' must be greater than 1 to satisfy the three mentioned objectives. In case of occupational exposures, a range of values between 1.2 and 1.8 seems reasonable (Schneider et al., 1997).

A.4. Cost-effectiveness analysis

(A18) Strictly speaking, cost-effectiveness analysis is not an optimisation technique but a method to eliminate non-cost-effective options from a set of options, and to rank and compare the remaining cost-effective options (ICRP, 1989). The basic principle of the method is to first characterise each protective option with its protective cost and the corresponding residual collective dose. The next step is selection of the cost-effective options, i.e. those for which there is no alternative solution allowing one to attain the same residual collective dose at a lower protection cost or the same level of protective cost with a lower residual collective dose.

(A19) This process can be illustrated in a simple way (Fig. A4). Each option is represented by a dot and all options that are cost-effective belong to the cost-effectiveness curve. For example, Option A allows one to reach a residual level of exposure at a lower cost than Option E, and Option C gives a lower residual collective dose for

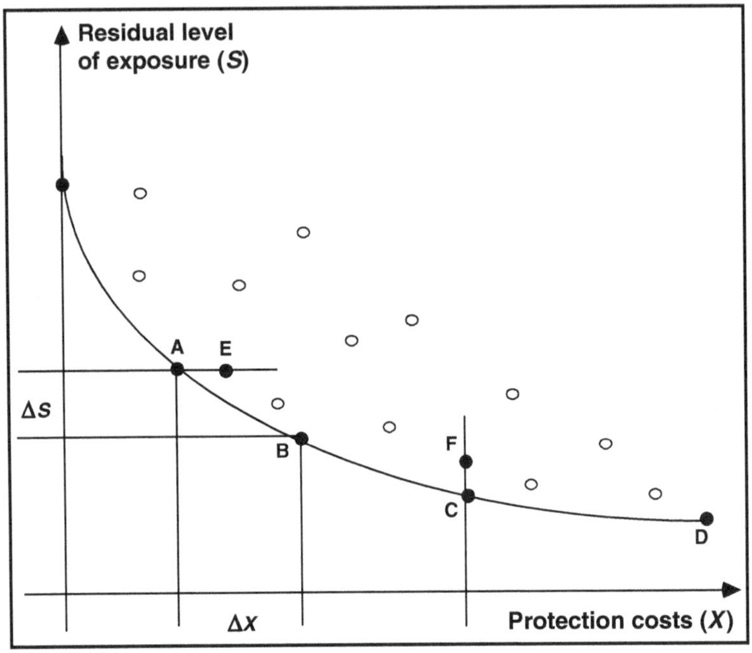

Fig. A4. Cost-effectiveness analysis.

the same cost as Option E. All the options that are not on the curve are not cost-effective and must be discarded from further consideration.

(A20) Formally, cost-effectiveness analysis relies on analysis of the 'marginal cost' of each protective option, which has to be compared with the immediately less or more expensive options. If a small additional cost leads to much higher effectiveness in terms of risk reduction, the option is more cost-effective. Finally, each cost-effective option can be characterised by the increase in cost from one option to the next [ΔX] and the corresponding decrease in collective dose [ΔS]. The quotient [$\Delta X/\Delta S$] is called the cost-effectiveness ratio, and this provides a basis for ranking between various options. The best option is the option with the ratio equal to or just below the monetary value of the man sievert selected as reference criteria for the particular exposure situation.

(A21) Nevertheless, determination of the cost-effectiveness curve, as well as the corresponding cost-effectiveness ratios, does not provide any basis for selection of the optimum option. This is done by introduction of a reference value for the cost-effectiveness ratio, i.e. the monetary value of the man-Sievert as it is defined above in relation to the cost–benefit analysis.

A.5. Multi-attribute utility analysis

(A22) When the relevant attributes to characterise the exposure situation, other than the radiation detriment and the protection costs, are numerous and/or difficult to quantify in monetary terms but nevertheless quantifiable in other criteria or able to be ranked in a qualitative way, it may be more appropriate to use multi-attribute utility analysis (MAUA) (ICRP, 1989).

(A23) The basic principle of this technique is to build a scoring scheme (or multi-attribute utility function) for each option on the basis of all the relevant criteria characterising the situation (i.e. cost of protective option, collective dose, individual dose, spread of the exposure in time and space, perception of the level of risk, etc.). Identification of the different protective options being the first step of the MAUA, consists of defining the relevant criteria for the specific decision process. Then, each protective option has to be evaluated according to the various criteria (either quantitatively or qualitatively). Due to the variety of criteria, the options are ranked differently for each criterion. The next step consists of assigning weighting factors for each criterion, expressing the relative importance associated with each of these criteria. It should be noted that this is the most important and often difficult step in a MAUA. Nevertheless, several techniques exist to derive this set of values, and whatever method is used, the choice of weighting factors should be justified.

(A24) Formally, the last step consists of qualifying each option by its total utility (U_i), calculated as follows:

$$U_i = \sum k_j u_{j,i}$$

where i is the index of option; j is the index of criterion; k_j is the weighting factor expressing the relative importance of each criterion j (normalised: $\sum k_j = 1$); and $u_{j,i}$ is the single utility for criterion j.

(A25) It is important to note that the single utility associated with each criterion can be defined either as a linear function of the value expressing the criterion, or as non-linear in order to incorporate the preferences of the decision makers into the analysis. For example, it is possible to define utility functions including risk aversion according to the level of individual exposures.

(A26) Finally, the protective option that leads to the highest total utility will be selected. As most weighting factors generally rely on the value judgments of decision makers, it is highly recommended that one should perform a sensitivity analysis according to different sets of weighting factors to test the 'robustness' of the results.

A.6. Conclusion

(A27) For many exposure situations, the use of decision-aiding techniques is an effective means to formalise and quantify selection of the best option in the optimisation process. The selection of a particular technique is mainly driven by the type of input data available and willingness to present the various attributes characterising the situation in the final results. In this respect, it is obvious that multi-attribute approaches are much more adapted to situations with conflicting attributes and decision makers' points of views. However, it is important to keep in mind that it is also possible to cope with many attributes in a cost–benefit framework provided that the weighting process related to the incorporated attributes is clearly elicited and a broad sensitivity analysis is performed to confirm the results.

(A28) Finally, it should be emphasised that difficulty with the implementation of decision-aiding techniques is not due to their intrinsic complexity but due to the complexity of the situations for which decisions have to be made with multiple attributes, protective options, and value judgment. In fact, whatever the decision-aiding technique to be implemented, cost-effectiveness, cost–benefit, or multi-attribute analysis, the major obstacles are in the delineation of the relevant attributes, the gathering of adequate data, and the integration of uncertainties and value judgments in a quantitative way.

REFERENCES

Beierle, 2002. Democracy in Practice – Public Participation in Environmental Decisions. Resources for the Future. Washington DC.

IAEA, 1991. Safety Culture. INSAG 4. International Atomic Energy Agency, Vienna, Austria.

IAEA, 2002. Optimisation of Radiation Protection in the Control of Occupational Exposure. Safety Report Series no. 21. International Atomic Energy Agency, Vienna, Austria.

ICRP, 1955. Recommendations of the International Commission on Radiological Protection (Revised December 1, 1954). Br. J. Radiol. (Suppl. 6).

ICRP, 1959. Recommendations of the International Commission on Radiological Protection. ICRP Publication 1, Pergamon Press, Oxford.

ICRP, 1966a. The Evaluation of Risks from Radiations. ICRP Publication 8, Pergamon Press, Oxford.

ICRP, 1966b. Recommendations of the International Commission on Radiological Protection. ICRP Publication 9, Pergamon Press, Oxford.

ICRP, 1973. Implications of Commission Recommendations that Doses be Kept as Low as Readily Achievable. ICRP Publication 22, Pergamon Press, Oxford.

ICRP, 1977. Recommendations of the International Commission on Radiological Protection. ICRP Publication 26. Ann. ICRP 1 (3).

ICRP, 1983. Cost–benefit analysis in the optimisation of radiological protection. ICRP Publication 37, Ann. ICRP 10 (2/3).

ICRP, 1988. Optimisation and decision-making in radiological protection. ICRP Publication 55, Ann. ICRP 20 (1).

ICRP, 1989. Optimization and decision-making in radiological protection, ICRP Publication 55, Ann. ICRP 20 (1).

ICRP, 1991. Recommendations of the International Commission on Radiological Protection. ICRP Publication 60, Ann. ICRP 21 (1–3).

ICRP, 1993. Principles for intervention for protection of the public in a radiological emergency. ICRP Publication 63, Ann. ICRP 22 (4).

ICRP, 1997. Radiological protection policy for the disposal of radioactive waste. ICRP Publication 77, Ann. ICRP 27 (Suppl.).

ICRP, 1998. General principles for the radiation protection of workers. ICRP Publication 75, Ann. ICRP 27 (1).

ICRP, 1999. Protection of the public in situations of prolonged radiation exposure. ICRP Publication 82, Ann. ICRP 29 (1–2).

IPSN, 2002. Collective Dose: Indications and Contraindications. Report from a Working Group. EDP Sciences, Paris, France.

Lochard, J., Lefaure, J., Schieber, C., Schneider, T., 1996. A model for the determination of monetary values of the man-Sievert. J. Radiol. Prot. 16, 201–204.

NEA, 1997. Work Management in the Nuclear Power Industry. Prepared for the NEA Committee on Radiation Protection and Public Health by the ISOE Expert Group on the Impact of Work Management on Occupational Exposure. OECD/Nuclear Energy Agency.

NEA, 1998. The Societal Aspects of Decision Making in Complex Radiological Situations. Proceedings of an International Workshop, Villigen, Switzerland, 13–15 January 1998. Organisation for Economic Co-operation and Development/Nuclear Energy Agency, Paris, France.

NEA, 2001. Better Integration of Radiation Protection in Modern Society. Proceedings of an International Workshop, Villigen, Switzerland, 23–25 January 2001. Organisation for Economic Co-operation and Development/Nuclear Energy Agency, Paris, France.

NEA, 2004. Stakeholder Participation in Decision Making Involving Radiation-exploring Processes and Implications. Proceedings of an International Workshop, Villigen, Switzerland, 21–23 October 2003. Organisation for Economic Co-operation and Development/Nuclear Energy Agency, Paris, France.

Schneider, T., Schieber, C., Eeckhoudt, L., Gollier, C., 1997. Economics of radiation protection: equity considerations. Theo. Decis. 43, 241–251.

Stokell, P., Croft, J., Lochard, J., Lombard, J., 1991. ALARA: from Theory towards Practice. Report EUR 13796 EN. Commission of the European Communities, Luxembourg.
UN, 1992. Report of the United Nations Conference in the Human Environment, Stockholm, 5–16 June 1992, United Nations Publication Sales No. E.73.II.A14 and corrigendum (Chap. 1).

he ICRP

rnational Commission

the primary body in protection against ionising
non-governmental organisation created by the
c benefit the science of radiological protection.
against the risks associated with ionishing ra-
stry and nuclear enterprises, and from naturally
d four times each year on behalf of the ICRP as
age of a specific subject area.

ars so that they are kept up to date on the latest
to acquire a complete set of ICRP reports and
arately for those individuals and organizations
ler through your bookseller, subscription agent,

ittees on: radiation effects, doses from radiation
ndations, and protection of the environment, all
consists of twelve members and a Chairman.
ical doctors dominate the current membership;

prepare its reports. A Task Group is usually
ajority of specialists from outside ICRP. Thus,
us fields of radiological protection. At any one
e work of ICRP. The Task Groups are assigned
are reviewed and finally approved by the Main
the ICRP.

(SSI), SE-171 16 Stockholm, Sweden

weden; e-mail: scient.secretary@icrp.org

FSC
www.fsc.org
MIX
Papier | Fördert gute Waldnutzung
FSC® C083411

Kolibri 360 GmbH
Ferdinand-Jühlke-Straße 7
99095 Erfurt, Deutschland
produktsicherheit@kolibri360.de